Human Interphase Chromosomes

Yuri B. Yurov • Svetlana G. Vorsanova
Ivan Y. Iourov
Editors

Human Interphase Chromosomes

Biomedical Aspects

Springer

Editors
Yuri B. Yurov
Mental Health Research Center
Russian Academy of Medical Sciences
Moscow, Russia

Institute of Pediatrics and Children Surgery
Ministry of Health
Moscow, Russia

Moscow City University of Psychology
and Education
Moscow, Russia

Ivan Y. Iourov
Mental Health Research Center
Russian Academy of Medical Sciences
Moscow, Russia

Institute of Pediatrics and Children Surgery
Ministry of Health
Moscow, Russia

Svetlana G. Vorsanova
Institute of Pediatrics and Children Surgery
Ministry of Health
Moscow, Russia

Mental Health Research Center
Russian Academy of Medical Sciences
Moscow, Russia

Moscow City University of Psychology
and Education
Moscow, Russia

ISBN 978-1-4614-6557-7 ISBN 978-1-4614-6558-4 (eBook)
DOI 10.1007/978-1-4614-6558-4
Springer New York Heidelberg Dordrecht London

Library of Congress Control Number: 2013934476

Printed on acid-free paper

Springer is part of Springer Science+Business Media (www.springer.com)

We dedicate this work to the memory of Ilia V. Soloviev, who will not be forgotten. He was a talented young researcher and a pioneer of molecular cytogenetics and of genome and chromosome research. His prodigious work and original ideas have formed our current research directions.

Dr. Ilia V. Soloviev

Preface

The study of human chromosomes in the context of an interphase nucleus is biologically most meaningful for understanding eukaryotic DNA expression and reproduction inasmuch as the interphase comprises essential periods of normal cell activity. To determine the architectural organization of chromosomes inside the nuclear space is thereby important for understanding how the genome functions during the cell cycle. Moreover, variations in chromosome number and structure in humans, who possess more than 200 types of cells, the majority of which are usually in interphase, cannot be properly addressed without using interphase cytogenetics (an umbrella term covering techniques for analysis of interphase chromosomes). The latter is often viewed as an esoteric discipline that only concerns a few specialists trying to implement single-cell approaches to genome biology and medicine. However, studying interphase chromosomes is relevant to numerous fields of life sciences, including, but not limited to, molecular and cell biology, biomedicine, genetics (including medical genetics), neuroscience, evolution, oncology, and genomics.

The beginning of experimental interphase cytogenetics can be attributed to significant advances in human molecular genetics and cytogenetics. As the consequence of experimental and theoretical research at the interface between cellular and molecular levels of chromosomal organization and function, high-resolution techniques for chromosomal analysis (molecular cytogenetic techniques) have become available. Molecular cytogenetics is a branch of biomedical sciences that explores chromosomes at molecular and single-cell resolutions at all stages of the cell cycle. It also comprises the techniques that operate with either the entire genome or specific DNA sequences to analyze genomic structural and functional variations at the chromosomal level. In the postgenomic era, molecular cytogenetics has appreciably transformed and has given rise to a new field of genomics, called cytogenomics. As a result, new opportunities have emerged for analysis of human interphase chromosomes in almost all cell types and states at unprecedented resolution. In this volume, we have attempted to provide an overview of current developments in the study of human interphase chromosomes with special attention to available molecular cytogenetic technologies for basic and clinical chromosome research.

The main body of the book is composed of 12 chapters. Chapter 1 (by Prof. Y.B. Yurov, Prof. S.G. Vorsanova, and Prof. I.Y. Iourov) is devoted to the basics of interphase molecular cytogenetics and cytogenomics in historical perspective. Chapter 2 (by Prof. J. Bridger and associates) considers contemporary views on interphase chromosome behavior in normal and diseased cells. The relationship between nuclear architecture and occurrence of chromosome aberrations is discussed in Chap. 3 (by Prof. G. Folle and Drs. Tomaso, Lafon-Hughes, and Liddle). The role of nuclear chromosome positioning, chromatin organization sensitivity to environmental exposures, genetic damage in metaphase and interphase, DNA replication and chromatin remodeling, and their involvements in the generation and localization of primary genetic damage are discussed. The unique possibility to visualize the interphase chromosome of the human brain and analyses of chromosome (genome) instability in postmitotic neuronal and glial cells are discussed in Chap. 4 (by Prof. I.Y. Iourov, Prof. S.G. Vorsanova, and Prof. Y.B. Yurov). In this chapter, mosaic aneuploidy is defined as a new feature of the normal human brain; increased chromosome instability in the developing and adult human brain is shown to be associated with neurodevelopmental and neurodegenerative genetic brain disorders (autism, schizophrenia, ataxia-telangiectasia, Alzheimer's disease); and interphase molecular cytogenetics is demonstrated to be the way for future studies of somatic genome instability and etiology (and pathogenesis) of genetic brain diseases. Taking into account the increased interest in somatic cell evolution mediated by genome alteration and its clinical significance, Dr. J. Stevens and Prof. H. Heng (Chap. 5) review mechanisms of chromosome fragmentation and premature chromosome condensation. They also discuss the mechanisms and definition of premature chromosome condensation and its applications to basic and clinical research. Chapter 6 (by Prof. E. Volpi) reviews the association between pathology, large-scale chromatin organization, and nuclear architecture in an enigmatic chromosome instability syndrome (ICF syndrome: a rare epigenetic disorder caused by autosomal recessive mutations, often fatal in childhood). Chapter 7 (by Prof. D. Griffin and Drs. Fonseka, Tempest, Thornhill, and Ioannou) overviews interphase cytogenetics of human embryos, highlighting the progress and contentious pitfalls that it encounters. Because interphase cytogenetics has important applications in prenatal medicine, other chapters outlined in this book pave the way for a range of exciting new studies that, potentially, might emerge on human embryos and show FISH as a still useful tool for rapid, low-cost, and robust cell-by-cell information. Chapter 8 (by Drs. O.S. Mudrak, L. Solovjeva, and V. Chagin) describes experimental data of studies dedicated to human spermatozoa and discusses the implications of sperm chromosome organization for male reproductive health. Chromosomes in human sperm nuclei adopt a hierarchy of structures from protamine toroids (the elementary units of DNA packaging) to the higher-order organization (chromosome territories), suggesting that chromatin organization in sperm may have functional significance. The intention of Chap. 9 (by Prof. I.Y. Iourov, Dr. T. Liehr, Prof. S. Vorsanova, and Prof. Y. Yurov) is to present the basics of interphase chromosome-specific multicolor banding (ICS-MCB) and to list its applications in different biomedical fields. Chapter 10 (by Dr. T. Liehr and his collaborators) is focused on technical limitations in biomedical research of interphase chromosomes in their integrity. To overcome these limitations, the authors have proposed a new technology based on three-dimensional suspension fluorescence in situ hybridization (3D S-FISH) with microdissection-based engineered DNA probes and multicolor chromosome banding (MCB). Chapter 11 (by Prof. S.G. Vorsanova and her collaborators) describes technological aspects and numerous approaches of interphase molecular cytogenetic, which are all useful for chromosomal analysis in almost all human cell types. Regardless of numerous technological difficulties encountered during studying human interphase chromosomes in health and disease, molecular cytogenetics or cytogenomics ("chromosomics") does provide for high-resolution single-cell analysis of genome organization, structure, and behavior at all stages of the cell cycle. Finally, the editors provide a list of references to websites containing regularly updated information on molecular cytogenetics and cytogenomics, including useful links to relevant websites (see Appendix).

Moscow, Russia

Yuri B. Yurov
Svetlana G. Vorsanova
Ivan Y. Iourov

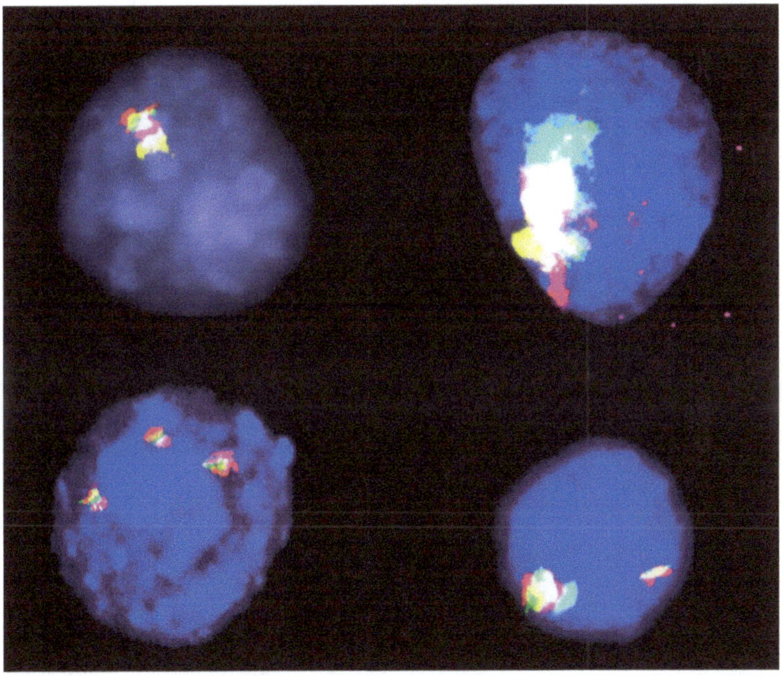

Molecular cytogenetic analyses of the developing and adult human brain by ICS-MCB. Upper row (*left*) - loss of chromosome 16 (monosomy) in a cell isolated from the prefrontal cortex of the normal human brain; (*right*) - loss of chromosome 1 (monosomy) in a cell isolated from the prefrontal cortex of the schizophrenia brain. Bottom row (*left*) - gain of chromosome 21 (trisomy) in a cell isolated from the prefrontal cortex of an Alzheimer's disease brain; (*right*) - chromosome instability in the cerebellum of the ataxia-telangiectasia brain manifesting as the presence of normal and a rearranged chromosome14 order (14)(14pter->14q12:)

Contents

Contributors

Joanna M. Bridger Laboratory for Nuclear and Genomic Health, CCCB, Brunel University, Kingston Lane, West London, UK

Samarth Bhatt Jena University Hospital, Institute of Human Genetics, Friedrich Schiller University, Jena, Germany

Ural Bikkul Laboratory of Cellular Gerontology, CCCB, Brunel University, Kingston Lane, West London, UK

Laboratory for Nuclear and Genomic Health, Centre for Cell and Chromosome Biology (CCCB), Brunel University, Kingston Lane, West London, UK

Gemma Bourne Laboratory of Cellular Gerontology, Centre for Cell and Chromosome Biology (CCCB), Brunel University, Kingston Lane, West London, UK

Vadim O. Chagin Institute of Cytology, Russian Academy of Sciences, St. Petersburg, Russia

Darren K. Griffin School of Biosciences, University of Kent, Canterbury, UK

María Vittoria Di Tomaso Department of Genetics, Instituto de Investigaciones Biológicas Clemente Estable (IIBCE), Montevideo, Uruguay

Christopher H. Eskiw Laboratory of Molecular Gerontology, Centre for Cell and Chromosome Biology (CCCB), Brunel University, Kingston Lane, West London, UK

Gustavo A. Folle Department of Genetics, Instituto de Investigaciones Biológicas Clemente Estable (IIBCE), Montevideo, Uruguay

Gothami Fonseka School of Biosciences, University of Kent, Canterbury, UK

Mai Hassan Ahmed Laboratory for Nuclear and Genomic Health, CCCB, Brunel University, Kingston Lane, West London, UK

Henry H.Q. Heng Center for Molecular Medicine and Genetics, Wayne State University School of Medicine, Detroit, MI, USA

Department of Pathology, Wayne State University School of Medicine, Detroit, MI, USA

Karmanos Cancer Institute, Detroit, MI, USA

Dimitris Ioannou Department of Human and Molecular Genetics, Florida International University, Miami, FL, USA

Ivan Y. Iourov Mental Health Research Center, Russian Academy of Medical Sciences, Moscow, Russia

Institute of Pediatrics and Children Surgery, Ministry of Health, Moscow, Russia

Tatyana Karamysheva Jena University Hospital, Institute of Human Genetics, Friedrich Schiller University, Jena, Germany

Laboratory of Morphology and Function of Cell structure, Institute of Cytology and Genetics, Russian Academy of Sciences, Siberian Branch, Novosibirsk, Russian Federation

Ian R. Kill Laboratory of Cellular Gerontology, Centre for Cell and Chromosome Biology (CCCB), Brunel University, Kingston Lane, West London, UK

Elisabeth Klein Jena University Hospital, Institute of Human Genetics, Friedrich Schiller University, Jena, Germany

Laura Lafon-Hughes Department of Genetics, Instituto de Investigaciones Biológicas Clemente Estable (IIBCE), Montevideo, Uruguay

Pablo Liddle Department of Genetics, Instituto de Investigaciones Biológicas Clemente Estable (IIBCE), Montevideo, Uruguay

Thomas Liehr Jena University Hospital, Institute of Human Genetics, Friedrich Schiller University, Jena, Germany

Marina Manvelyan Jena University Hospital, Institute of Human Genetics, Friedrich Schiller University, Jena, Germany

Research Center of Maternal and Child Health Protection, Yerevan, Armenia

Hasmik Mkrtchyan Jena University Hospital, Institute of Human Genetics, Friedrich Schiller University, Jena, Germany

Centre of Medical Genetics and Primary Health Care, Department of Medical Genetics, Yerevan, Armenia

Catherine Moir Laboratory of Molecular Gerontology, Centre for Cell and Chromosome Biology (CCCB), Brunel University, Kingston Lane, West London, UK

Olga S. Mudrak Institute of Cytology, Russian Academy of Sciences, St. Petersburg, Russia

Liudmila V. Solovjeva Institute of Cytology, Russian Academy of Sciences, St. Petersburg, Russia

Joshua B. Stevens Center for Molecular Medicine and Genetics, Wayne State University School of Medicine, Detroit, MI, USA

Helen G. Tempest Department of Human and Molecular Genetics, Florida International University, Miami, FL, USA

Alan R. Thornhill The London Bridge Fertility, Gynaecology and Genetics Centre, London, UK

Sabrina Tosi Leukaemia and Chromosome Research Laboratory, CCCB, Biosciences, Brunel University, Kingston Lane, Uxbridge, UK

Emanuela V. Volpi Wellcome Trust Centre for Human Genetics, University of Oxford, Oxford, UK

Svetlana G. Vorsanova Mental Health Research Center, Russian Academy of Medical Sciences, Moscow, Russia

Institute of Pediatrics and Children Surgery, Ministry of Health, Moscow, Russia

Moscow City University of Psychology and Education, Moscow, Russia

Yuri B. Yurov Mental Health Research Center, Russian Academy of Medical Sciences, Moscow, Russia

Institute of Pediatrics and Children Surgery, Ministry of Health, Moscow, Russia

Moscow City University of Psychology and Education, Moscow, Russia

Chapter 1
Introduction to Interphase Molecular Cytogenetics

Yuri B. Yurov, Svetlana G. Vorsanova, and Ivan Y. Iourov

Abstract The history of interphase cytogenetics can be traced back to the pioneering works on descriptions of intracellular compartments dated at the end of the nineteenth century. However, it was not until the development of molecular cytogenetic techniques that the direct analysis of human interphase chromosomes began. During the past three decades, tremendous efforts have been made toward the elucidation of how the cellular genome is organized at molecular and supramolecular (chromatin and chromosomal) levels. As a result, we do possess powerful molecular cytogenetic technologies for diagnosing chromosome abnormalities in interphase and studying chromosome number, structure, and behavior variations in single cells at molecular resolutions through the entire cell cycle. Using several seminal reviews as milestones, it was possible to show the development of interphase (molecular) cytogenetics in historical perspective. As one can notice, the main achievements in studying interphase chromosomes were made because of technological developments in molecular cytogenetics. Therefore, the present introduction to interphase molecular cytogenetics is not only limited to listing changing of concepts in studying interphase chromosomal architecture and molecular cytogenetic diagnosis, but also briefly describes the technological basis of this dynamically developing biomedical field.

Y.B. Yurov (✉) • S.G. Vorsanova
Mental Health Research Center, Russian Academy of Medical Sciences,
Moscow 119152, Russia

Institute of Pediatrics and Children Surgery, Ministry of Health, Moscow, Russia

Moscow City University of Psychology and Education, Moscow, Russia
e-mail: y_yurov@yahoo.com

I.Y. Iourov (✉)
Mental Health Research Center, Russian Academy of Medical Sciences,
Moscow 119152, Russia

Institute of Pediatrics and Children Surgery, Ministry of Health, Moscow, Russia
e-mail: ivan.iourov@gmail.com

Y.B. Yurov et al. (eds.), *Human Interphase Chromosomes: Biomedical Aspects*,
DOI 10.1007/978-1-4614-6558-4_1, © Springer Science+Business Media, LLC 2013

> Eukaryotic DNA expresses and reproduces itself only in the context of an interphase nucleus. It is therefore biologically most meaningful to understand chromosome organization in this state
>
> (Laura Manuelidis 1990).

It is hard to disagree with Professor Manuelidis that studying interphase chromosomes is the only way toward real understanding of the genome organization and variation. Because the interphase comprises essential periods of normal cell activity, one has to refer to molecular interphase cytogenetics for uncovering somatic genome variations in different tissues and analyzing the cellular genome organization. Originating from the pioneering works of Flemming (in 1882), Rabl (in 1885), Waldeyer (in 1888), and Boveri (in 1909) (for more details and references see Smeets 2004; Foster and Bridger 2005; Cremer and Cremer 2006), studies dedicated to the structural and functional organization of genetic material at the chromosomal level in interphase remained almost completely theoretical (Comings 1968, 1980; Vogel and Schroeder 1974; Manuelidis 1990), until molecular cytogenetic methods forming a firm technical basis for high-resolution detection of chromosomal regions in metaphase and interphase cells became available (Trask 2002; Levsky and Singer 2003; Claussen 2005; Iourov et al. 2006a, 2008b, 2012; Serakinci and Kolvraa 2009; Vorsanova et al. 2010a; Pajor et al. 2012). Actually, the essential knowledge about genome behavior in the interphase (chromosome architecture, somatic genome variations, etc.) was gained during the past three decades, which can be fairly called "the era of molecular cytogenetics" from a cytogeneticist's point of view. Moreover, the wide use of postgenomic molecular cytogenetic techniques based on data acquired by an extended number of human genome studies has given rise to a new field of genomics: cytogenomics (Smeets 2004; Iourov et al. 2008a, 2012; Vanneste et al. 2012). Therefore, to provide an introduction to interphase molecular cytogenetics and cytogenomics in historical perspective, it is mandatory to address (i) concepts in understanding chromosome structural and functional organization in the interphase (genome organization and variation) and (ii) technological milestones in interphase cytogenetics.

The history of human cytogenetics (including molecular cytogenetics) has been repeatedly addressed by a series of high-quality reviews (Trask 2002; Levsky and Singer 2003; Smeets 2004; Foster and Bridger 2005; Gersen and Keagle 2005; Cremer and Cremer 2006; Gartler 2006; Serakinci and Kolvraa 2009). In this instance, we have preferred to focus more on changes of concepts in studying interphase chromosomal architecture and molecular cytogenetic diagnosis that have occurred through the history of interphase cytogenetics rather than to give the well-known timeline of cytogenetics history.

The complexity of internal order in the interphase nucleus and arrangement of all its compartments has been long acknowledged. Furthermore, technological difficulties in analysis of intranuclear organelles seem to be the reason why the nucleus has been ironically termed "the black box" (van Driel et al. 1991). Nonetheless, earlier studies (reported in the late nineteenth and early twentieth centuries) have proposed a hypothesis suggesting that interphase chromosomes are likely to occupy more-or-less distinct territories (for more details, see Cremer and Cremer 2006). Formally, in the middle of the twentieth century, there have already been studies roughly depicting interphase chromosomes or, more precisely, their specific behavior (i.e., visualization of facultative heterochromatin or sex chromatin, termed later the Barr body) (Barr and Bertram 1949). In the beginning of the second half of the twentieth century, the hypothesis suggesting the existence of chromosome territories was left behind as evaluations of heterochromatin by chromosome staining techniques and modeling of interphase chromosome behavior according to metaphase chromosomal analyses suggested that heterochromatic regions should be the only chromosomal regions that hold nonrandom intranuclear positions, whereas euchromatic chromosomal regions are likely to fill nearly completely the nuclear space apart from those peculiar territories occupied by heterochromatin and other non-DNA nuclear components (Comings 1968, 1980; Vogel and Schroeder 1974). Although a rationale for the internal order of chromatin arrangement within the nucleus was postulated (Comings 1968; Vogel and Schroeder 1974), some questions concerning the way that genome is processed and chromosomes are arranged within the nuclear space remained unanswered (for more details, see Comings 1980; Foster and Bridger 2005; Cremer and Cremer 2006). The introduction of in situ hybridization to basic chromosome biology (interphase chromosomal analysis) has changed the main concept in nuclear genome organization (Manuelidis 1990). However, suggestions about heterochromatin arrangement in interphase nuclei proposed by Comings (Comings 1968) have remained actual. Fluorescence in situ hybridization (FISH) has provided for numerous possibilities to uncover chromosome architecture, and its applications resulted in a significant reevaluation of concepts in nuclear genome organization (Leitch 2000; Foster and Bridger 2005; Cremer and Cremer 2006). Nonrandomness in interphase chromosome architecture (the existence of chromosome territories) has been proposed as the central point in speculations about the driving force of the cellular genome organization (Foster and Bridger 2005; Cremer and Cremer 2006; Rajapakse and Groudine 2011). Considering the data on the plasticity of chromosomes and its associations with regulation of critical intranuclear processes (chromosomics) (Claussen 2005), it was postulated that chromosome architecture in interphase not only possesses a specific internal order, but also is the last hierarchical level of cellular genome organization and is the final step of genome behavior modulation (Cremer and Cremer 2006; Rajapakse and Groudine 2011). Currently, interphase chromosome architecture is suggested to play an important role in numerous critical nuclear processes: modulation of transcriptional activity through chromatin organization; regulation of developmental pathways; chromosomal DNA replication, genome maintenance, and DNA repair; promoting somatic interchromosomal rearrangements; programmed cell death; and determination of the genomic landscape in senescent cells (mechanism for aging at the cellular level) (Leitch 2000; Foster and Bridger 2005; Misteli and Soutoglou 2009; Rouquette et al. 2010; Rajapakse and Groudine 2011). Additionally, it is generally accepted that interphase chromosome associations (somatic pairing) seem to be involved in regulation of transcriptional activity within specific chromosomal regions, including imprinted genomic loci, and are probably involved in the aforementioned nuclear processes as well (Göndör and Ohlson 2009; Dostie and Bickmore 2012). All these data become available because of technical breakthroughs in molecular cytogenetics and successful combinations of interphase (molecular) cytogenetic techniques and specific microscopy/imaging approaches and more recent (on-chip or "on-beads") technologies probing chromatin states and interactions (for reviews, see Rouquette et al. 2010 and Dostie and Bickmore 2012).

In a similar way as for interphase chromosomes architecture, understanding of somatic genome variations was long hindered by a dogma, which is briefly formulated as follows: "with few exceptions, all the cells of an organism share identical genomes." Currently, there is a plethora of data that directly show all humans to be somatic mosaics. Consequently, studying somatic genome variations using interphase molecular cytogenetic techniques has become an appreciable part of biomedical research (for more details, see Iourov et al. 2006a, b, 2010, 2012; Vorsanova et al. 2010b). The latter has been the result of developments in interphase molecular cytogenetics [i.e., elaborations of FISH and FISH-based approaches as well as single-cell whole-genome scan approaches based on array comparative genome hybridization (array CGH)]. Here again, the technological achievements in molecular cytogenetics were the key to interphase cytogenetics relevant to somatic cell genetics and genomics or, more precisely, cytogenomics. Additionally, these achievements allowed the development of interphase molecular cytogenetic diagnosis, which has become an important part of the clinical workup for patients suffering from congenital malformations, neurobehavioral diseases, reproductive problems, and cancer (Vorsanova et al. 1991, 2010a; b; Tanke et al. 1995; Ried 1998; Levsky and Singer 2003; Iourov et al. 2008b). Interphase FISH is also found applicable in prenatal diagnosis of chromosome abnormalities (Soloviev et al. 1995; Bui et al. 2002). In addition, numerous targeted interphase FISH assays are proven effective for detecting specific chromosome rearrangements in preimplantation and oncocytogenetic diagnosis (see Chap. 11).

The technological basis of FISH (or in situ hybridization) was formed during the 1970s and 1980s and included the development of DNA–DNA (DNA–RNA/RNA–DNA) hybridization protocols, labeling of DNA methods (i.e., nick translation or similar methods), and elaboration of digital analysis and imaging protocols for fluorescence microscopy (for more details, see Tanke et al. 1995; Levsky and Singer 2003; Serakinci and Kolvraa 2009). In 1986, in situ hybridization (FISH) was demonstrated to be a valuable tool for molecular diagnosis of chromosomal abnormalities (Cremer et al. 1986; Pinkel et al. 1986; Vorsanova et al. 1986). That year is arbitrarily considered as the starting point of the molecular cytogenetic era or molecular cytogenetics. During the following 10 years, several important contributions were made to FISH-based technologies, resulting in the elaboration of CGH (Kallioniemi et al. 1992) and multicolor FISH (spectral karyotyping), allowing differential painting of each human homologous chromosome pair

(Schrock et al. 1996; Speicher et al. 1996). However, all these techniques were based on studying either the total DNA or metaphase chromosomes. It took another 10 years to make these approaches useful for studying genomes and chromosomes of interphase cells (for review, see Iourov et al. 2008b; Vorsanova et al. 2010a; Vanneste et al. 2012). In interphase molecular cytogenetics, there have been proposals to use FISH for studying specific chromosomal regions using chromosome-specific differentially labeled DNA probes that allowed multitarget interphase chromosomal analysis (Ried et al. 1992; Yurov et al. 1996). To increase the efficiency of interphase FISH, a combination of FISH and immunocytochemistry (immuno-FISH) has been proposed (for review, see Tanke et al. 2005). This approach has been subsequently found valuable for different areas of chromosome biology and has been used as a method of choice in studying somatic genome variations and interphase chromosome architecture. FISH-based approaches were also used for generating a protocol of interphase multicolor chromosome-specific banding, a method for analysis of interphase chromosomes in their integrity at molecular resolutions (Iourov et al. 2007; described in Chap. 9). All these approaches are described and discussed in more detail in Chap. 11. It is to be added that FISH can be used as a platform for automated interphase molecular cytogenetic analysis (Pajor et al. 2012). Finally, CGH has been used as a platform for an on-chip whole-genome analysis (array CGH) (reviewed in Trask 2002; Smeets 2004), which is considered to be the most powerful molecular cytogenetic technique for detection of constitutional chromosome imbalances. Recently, it has been shown that the whole-genome scan by array-CGH or array-CGH-based technologies is applicable to single cells, providing for an intriguing alternative to other interphase molecular cytogenetic techniques used for evaluations of somatic genome variations (for review, see Vanneste et al. 2012). However, because of natural limitations, the single-cell array-CGH whole-genome scan is hardly applicable to studying large cell populations. In conclusion, interphase molecular cytogenetics does possess the technological opportunities for studying specific interphase chromosomal loci, whole interphase chromosomes (at a resolution comparable to or even higher than that of metaphase analysis), and the whole cellular genome (Table 1.1).

Last, because all the high-quality research in the field of interphase molecular cytogenetics is almost impossible to reference in a book chapter, we would like to apologize to authors whose papers were not discussed (cited) owing to the introductory nature of this chapter. Fortunately, a great number of such articles are discussed and reviewed in the next chapters, which are focused on more specific areas of interphase cytogenetics. We hope that this book, dedicated to human interphase chromosomes and interphase molecular cytogenetics, will serve as a valuable addition to current biomedical literature and will be useful to all who perform research and those who teach in the fields of chromosome and genome biology, medical genetics, and related biomedical fields.

Table 1.1 Technical basis of interphase molecular cytogenetics

Platform	Technique	Description and resolution	Key references[a]
Fluorescence in situ hybridization (FISH)	I-FISH[b] painting of specific genomic loci	FISH on interphase nuclei using site-specific or chromosome enumeration probes; >50–100 kb	Ried (1998); Levsky and Singer (2003); Tanke et al. (2005); Iourov et al. (2008a); Vorsanova et al. (2010a); Pajor et al. (2012); Chap. 11 (this book)
	I-FISH painting of whole chromosomes	FISH on interphase nuclei using whole-chromosome probes; the length of a chromosome	
	ICS-MCB[c]	Multicolor chromosomal banding applied to interphase chromosomes; >2–5 Mb	
FISH + immunocytochemistry	Immuno-FISH	Combination of immunohistochemical protein detection and interphase FISH; resolution depends on DNA probes (as for I-FISH painting of specific genomic loci)	
Array comparative genome hybridization (CGH)	Single-cell array CGH—bacterial artificial chromosome (BAC) arrays Single-cell array CGH—oligoarrays Single-cell array CGH—SNP arrays	Array CGH protocols using either BAC or oligonucleotides as probes or a single-nucleotide polymorphism (SNP) array platform. As shown by single-cell studies, the method enables detection of genome imbalances >600 kb	Vanneste et al. (2012)

[a]Reviews only
[b]Interphase FISH
[c]Interphase chromosome-specific multicolor banding

References

Barr ML, Bertram EG (1949) A morphological distinction between neurones of the male and female, and the behaviour of the nucleolar satellite during accelerated nucleoprotein synthesis. Nature (Lond) 163:676–677

Bui TH, Blennow E, Nordenskjöld M (2002) Prenatal diagnosis: molecular genetics and cytogenetics. Best Pract Res Clin Obstet Gynaecol 16(5):629–643

Claussen U (2005) Chromosomics. Cytogenet Genome Res 111:101–106

Comings DE (1968) The rationale for an ordered arrangement of chromatin in the interphase nucleus. Am J Hum Genet 20(5):440–460

Comings DE (1980) Arrangement of chromatin in the nucleus. Hum Genet 53(2):131–143

Cremer T, Cremer C (2006) Rise, fall and resurrection of chromosome territories: a historical perspective. Part II. Fall and resurrection of chromosome territories during the 1950s to 1980s.

Part III. Chromosome territories and the functional nuclear architecture: experiments and models from the 1990s to the present. Eur J Histochem 50(4):223–272

Cremer T, Landegent J, Brückner A, Scholl HP, Schardin M, Hager HD et al (1986) Detection of chromosome aberrations in the human interphase nucleus by visualization of specific target DNAs with radioactive and non-radioactive in situ hybridization techniques: diagnosis of trisomy 18 with probe L1.84. Hum Genet 74(4):346–352

Dostie J, Bickmore WA (2012) Chromosome organization in the nucleus—charting new territory across the Hi-Cs. Curr Opin Genet Dev 22(2):125–131

Foster HA, Bridger JM (2005) The genome and the nucleus: a marriage made by evolution. Genome organisation and nuclear architecture. Chromosoma (Berl) 114(4):212–229

Gartler SM (2006) The chromosome number in humans: a brief history. Nat Rev Genet 7(8):655–660

Gersen SL, Keagle MB (2005) The principles of clinical cytogenetics, 2nd edn. Humana Press, Totowa, NJ

Göndör A, Ohlson R (2009) Chromosome crosstalk in three dimensions. Nature (Lond) 461:212–217

Iourov IY, Vorsanova SG, Yurov YB (2006a) Chromosomal variations in mammalian neuronal cells: known facts and attractive hypotheses. Int Rev Cytol 249:143–191

Iourov IY, Vorsanova SG, Yurov YB (2006b) Intercellular genomic (chromosomal) variations resulting in somatic mosaicism: mechanisms and consequences. Curr Genomics 7:435–446

Iourov IY, Liehr T, Vorsanova SG, Yurov YB (2007) Interphase chromosome-specific multicolor banding (ICS-MCB): a new tool for analysis of interphase chromosomes in their integrity. Biomol Eng 24:415–417

Iourov IY, Vorsanova SG, Yurov YB (2008a) Molecular cytogenetics and cytogenomics of brain diseases. Curr Genomics 9(7):452–465

Iourov IY, Vorsanova SG, Yurov YB (2008b) Recent patents on molecular cytogenetics. Recent Pat DNA Gene Seq 2:6–15

Iourov IY, Vorsanova SG, Yurov YB (2010) Somatic genome variations in health and disease. Curr Genomics 11(6):387–396

Iourov IY, Vorsanova SG, Yurov YB (2012) Single cell genomics of the brain: focus on neuronal diversity and neuropsychiatric diseases. Curr Genomics 13(6):477–488

Kallioniemi A, Kallioniemi OP, Sudar D, Rutovitz D, Gray JW, Waldman F, Pinkel D (1992) Comparative genomic hybridization for molecular cytogenetic analysis of solid tumors. Science 258(5083):818–821

Leitch AR (2000) Higher levels of organization in the interphase nucleus of cycling and differentiated cells. Microbiol Mol Biol Rev 64(1):138–152

Levsky JM, Singer RH (2003) Fluorescence in situ hybridization: past, present and future. J Cell Sci 116(pt 14):2833–2888

Manuelidis L (1990) A view of interphase chromosomes. Science 250(4987):1533–1540

Misteli T, Soutoglou E (2009) The emerging role of nuclear architecture in DNA repair and genome maintenance. Nat Rev Mol Cell Biol 10:243–254

Pajor G, Kajtár B, Pajor L, Alpár D (2012) State-of-the-art FISHing: automated analysis of cytogenetic aberrations in interphase nuclei. Cytometry A 81(8):649–663

Pinkel D, Straume T, Gray JW (1986) Cytogenetic analysis using quantitative, high-sensitivity, fluorescence hybridization. Proc Natl Acad Sci USA 83(9):2934–2938

Rajapakse I, Groudine M (2011) On emerging nuclear order. J Cell Biol 192(5):711–721

Ried T (1998) Interphase cytogenetics and its role in molecular diagnostics of solid tumors. Am J Pathol 152(2):325–327

Ried T, Landes G, Dackowski W, Klinger K, Ward DC (1992) Multicolor fluorescence in situ hybridization for the simultaneous detection of probe sets for chromosomes 13, 18, 21, X and Y in uncultured amniotic fluid cells. Hum Mol Genet 1(5):307–313

Rouquette J, Cremer C, Cremer T, Fakan S (2010) Functional nuclear architecture studied by microscopy: present and future. Int Rev Cell Mol Biol 282:1–90

Schrock E, du Manoir S, Veldman T, Schoell B, Weinberg J, Ferguson-Smith MA et al (1996) Multicolor spectral karyotyping of human chromosomes. Science 273:494–497

Serakinci N, Kolvraa S (2009) Molecular cytogenetic applications in diagnostics and research: an overview. In: Liehr T (ed) Fluorescence in situ hybridization (FISH): application guide. Springer, Berlin, pp 3–21

Smeets DF (2004) Historical prospective of human cytogenetics: from microscope to microarray. Clin Biochem 37(6):439–446

Soloviev IV, Yurov YB, Vorsanova SG, Fayet F, Roizes G, Malet P (1995) Prenatal diagnosis of trisomy 21 using interphase fluorescence in situ hybridization of postreplicated cells with site-specific cosmid and cosmid contig probes. Prenat Diagn 15:237–248

Speicher MR, Ballard GS, Ward DC (1996) Karyotyping human chromosomes by combinatorial multi-fluor FISH. Nat Genet 12:368–375

Tanke HJ, Florijn RJ, Wiegant J, Raap AK, Vrolijk J (1995) CCD microscopy and image analysis of cells and chromosomes stained by fluorescence in situ hybridization. Histochem J 27(1):4–14

Tanke HJ, Dirks RW, Raap T (2005) FISH and immunocytochemistry: towards visualising single target molecules in living cells. Curr Opin Biotechnol 16(1):49–54

Trask BJ (2002) Human cytogenetics: 46 chromosomes, 46 years and counting. Nat Rev Genet 3(10):769–778

van Driel R, Humbel B, de Jong L (1991) The nucleus: a black box being opened. J Cell Biochem 47(4):311–316

Vanneste E, Bittman L, Van der Aa N, Voet T, Vermeesch JR (2012) New array approaches to explore single cells genomes. Front Genet 3:44

Vogel F, Schroeder TM (1974) The internal order of the interphase nucleus. Humangenetik 25(4):265–297

Vorsanova SG, Yurov YB, Alexandrov IA, Demidova IA, Mitkevich SP, Tirskaya AF (1986) 18p-syndrome: an unusual case and diagnosis by in situ hybridization with chromosome 18-specific alphoid DNA sequence. Hum Genet 72:185–187

Vorsanova SG, Yurov YB, Deryagin GV, Soloviev IV, Bytenskaya GA (1991) Diagnosis of aneuploidy by in situ hybridization: analysis of interphase nuclei. Bull Exp Biol Med 112:413–415

Vorsanova SG, Yurov YB, Iourov IY (2010a) Human interphase chromosomes: a review of available molecular cytogenetic technologies. Mol Cytogenet 3:1

Vorsanova SG, Yurov YB, Soloviev IV, Iourov IY (2010b) Molecular cytogenetic diagnosis and somatic genome variations. Curr Genomics 11:440–446

Yurov YB, Soloviev IV, Vorsanova SG, Marcais B, Roizes G, Lewis R (1996) High resolution fluorescence in situ hybridization using cyanine and fluorescein dyes: ultra-rapid chromosome detection by directly fluorescently labeled alphoid DNA probes. Hum Genet 97:390–398

Chapter 2
Interphase Chromosome Behavior in Normal and Diseased Cells

Gemma Bourne, Catherine Moir, Ural Bikkul, Mai Hassan Ahmed, Ian R. Kill, Christopher H. Eskiw, Sabrina Tosi, and Joanna M. Bridger

Abstract Interphase chromosomes are nonrandomly positioned in the nuclei of normal cells. They occupy specific locations with respect to a radial distribution from the nuclear edge to the nuclear interior. Furthermore, there is some evidence that interphase chromosomes reproducibly have the same neighbors that can be involved in creating translocations which lead to cancer. Not only are chromosomes nonrandomly positioned but they are anchored to certain regions of the cell nucleus by cellular structures such as the nuclear lamina and the nucleolus. Global screening of the genome has identified both lamina-associated domains and nucleolar-associated domains. Increasingly, researchers are finding that interphase chromosomes are mislocalized in disease situations. The consequences of chromosome mislocalization are not yet that clear, but gene expression can be affected with interphase chromosomes being located in another compartment of the nucleus, changing

G. Bourne • I.R. Kill
Laboratory of Cellular Gerontology, Centre for Cell and Chromosome Biology (CCCB), Brunel University, Kingston Lane, West London UB8 3PH, UK

C. Moir • C.H. Eskiw
Laboratory of Molecular Gerontology, CCCB, Brunel University, Kingston Lane, West London UB8 3PH, UK

U. Bikkul
Laboratory of Cellular Gerontology, CCCB, Brunel University, Kingston Lane, West London UB8 3PH, UK

Laboratory for Nuclear and Genomic Health, CCCB, Brunel University, Kingston Lane, West London UB8 3PH, UK

M.H. Ahmed • J.M. Bridger (✉)
Laboratory for Nuclear and Genomic Health, CCCB, Brunel University, Kingston Lane, West London UB8 3PH, UK
e-mail: joanna.bridger@brunel.ac.uk

S. Tosi
Leukaemia and Chromosome Research Laboratory, CCCB, Biosciences, Brunel University, Kingston Lane, Uxbridge UB8 3PH, UK

Y.B. Yurov et al. (eds.), *Human Interphase Chromosomes: Biomedical Aspects*,
DOI 10.1007/978-1-4614-6558-4_2,

their interactions with nuclear structures. This chapter outlines how chromosomes behave in interphase nuclei and with what they interact. We discuss many examples of when chromosomes, and the genes housed upon them, change their location and behavior in disease situations such as cancer and the premature aging syndrome called Hutchinson–Gilford progeria syndrome. We also describe new findings whereby genes in the host are relocated and expressed after a parasitic infection.

Introduction

The eukaryotic cell nucleus is a highly complex organelle that contains the cell genome in the form of interphase chromosome territories. Interphase chromosomes are described thus because as they decondense, after cell division within the new daughter nuclei, they remain in and interact with a particular area of the nucleus, a so-called territory of the nucleus. Although there is some intermingling and 'coming together' of genes from different chromosomes, most of the body of one chromosome is maintained together, and thus to all intents and purposes chromosomes in interphase are separate entities. The interphase chromosomes are nonrandomly positioned in nuclei, revealing that there must be a high level of genome reorganization post mitosis to obtain individual chromosome territories in the right compartments of the nucleus (Fig. 2.1).

The structures within nuclei are intimately involved in organizing and positioning interphase chromosomes to allow the coordination of a wide range of functions centred around the genome, such as gene expression and silencing, splicing and processing, and DNA replication and DNA repair. These nuclear structures are all linked and are part of a functionally responsive cellular network (Starr 2009). Such architecture comprises the nuclear envelope with all its components: integral membrane proteins (IMPs) and the nuclear lamina, nucleoli, the nucleoskeleton, and a range of nuclear bodies (Foster and Bridger 2005). All these structures interact with and anchor interphase chromosomes. Misorganization or disruption of this nuclear architecture can lead to problems in regulating normal chromosome behavior, producing compromised cells with the possibility that diseases such as cancer or degenerative syndromes may arise.

Nuclear Structures

The most prominent subcompartment of the nucleus is the nucleolus. The nucleoli are where ribosomal RNAs are synthesized and processed, thereby providing a site for efficient assembly of ribosomal subunits. In humans the acrocentric chromosomes containing the ribosomal repeat genes are embedded in the nucleoli, providing a functional anchorage site for these genes and their chromosomes (Bridger et al. 1998). tRNA genes are also clustered at the nucleoli (Boisvert et al. 2007; Nemeth et al. 2010). Furthermore, other chromosomes that do not contain ribosomal DNA or tRNA genes are also associated with nucleoli (Bridger et al. 1998).

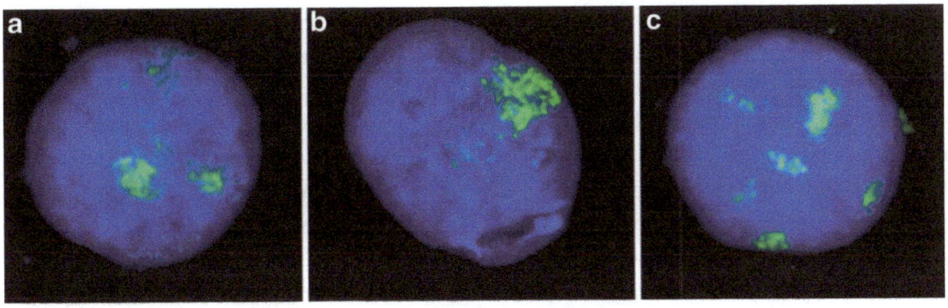

Fig. 2.1 Chromosome territories in interphase nuclei. Individual territories for human chromosome 10 (*green*) have been delineated using two-dimensional (2D) fluorescence in situ hybridization (FISH). (**a**) Normal immortalized human dermal fibroblast interphase nucleus. (**b**) Immortalized Hutchinson–Gilford progeria syndrome fibroblast nucleus. (**c**) Nucleus from a transformed cell derived from a breast tumor. Note that the nuclei in (**a**) and (**b**) contain only two territories whereas (**c**) displays many territories of chromosome 10 that are derived from ploidy and translocations

In genome-wide screens, two studies have revealed many sites throughout the genome that are anchored at the nucleoli, including chromosome 17 (van Koningsbruggen et al. 2010) and chromosome 19 (Nemeth et al. 2010). Both these chromosomes have been found to be interiorly located and associated with nucleoli in extracted nuclei (J. Bridger, unpublished data). These studies demonstrate that the nucleolus is a major player in anchoring and organizing chromosome territories in interphase nuclei.

The interior of a nucleus is thought to be more conducive to transcription of active genes, whereas a correlation has been shown between gene repression and positioning at the periphery (Zink et al. 2004). In yeast the nuclear periphery has been shown to consist of two distinct compartments: a region permissive to transcription near the nuclear pore complexes, and a repressive region that contains foci of silencing factors (Andrulis et al. 1998; Taddei et al. 2006). Components of the nuclear periphery, such as nucleoporins and lamin proteins, are thought to interact with repressors of transcription. For example, emerin interacts with the transcriptional repressors germ cell-less (GCL) and barrier to autointegration factor (BAF) (Holaska et al. 2003), and the nuclear envelope protein LAP2β interacts with HDAC3 to cause histone H4 deacetylation and gene repression (Somech et al. 2005). Genes can become anchored at the periphery of the cell, which affects their local chromatin environment. For example, genes that become tethered to the nucleoporin Nup2p are blocked from becoming heterochromatic and therefore remain active, whereas tethering of telomeres to other nucleoporins results in gene silencing (Ishii et al. 2002; Feuerbach et al. 2002).

The nuclear envelope is made up of the inner and outer nuclear membranes, which house nuclear pore complexes, and the nuclear lamina. The inner nuclear membrane, facing the nuclear interior, contains a large number of IMPs (Trinkle-Mulcahy and Lamond 2007; Gomez-Cavazos and Hetzer 2012). By proteomic analysis, at least 67 IMPs have been identified. The better known IMPs are lamin B receptor, lamin-associated polypeptides 1 and 2, emerin, MAN1, and nesprins (Schirmer et al. 2003). Many of these IMPs have chromatin/DNA-binding capabilities and are believed to be involved in anchoring interphase chromosomes to the nuclear periphery (Zuleger et al. 2011). Furthermore, the nuclear envelope components can be very different in different tissue types (Korfali et al. 2012), which could explain why different areas of the genome become localized to the nuclear envelope in different cell types (Peric-Hupkes et al. 2010).

The nuclear lamina, found subjacent to the nuclear membrane, consists of type V intermediate filament proteins nuclear lamins A, B, and C and is known as a structural scaffold under the nuclear envelope, which provides mechanical strength. There are three mammalian lamin genes: *LMNA*, *LMNB1*, and *LMNB2*, encoding at least six proteins. *LMNA* encodes four alternatively spliced mRNAs for lamin A, AΔ10, and C1 and C2 proteins, which are called A-type lamins. *LMNB1* encodes lamin B1, and the lamin B2 mRNA can be spliced to yield B2 or B3 proteins. The presence of lamins A and C is limited to differentiated cells; however, lamins B1 and B2 are expressed in all cell types both in adults and in embryos. Furthermore, expression of certain lamin proteins such as C2 is restricted to the testis and during meiosis, whereas lamin B3 exists only in oocytes and spermatozoa (Rodríguez and Eriksson 2010). Lamin proteins have DNA/chromatin-binding abilities but also bind to a number of the IMPs of the nuclear membrane. Thus, there are a plethora of sites at the nuclear periphery for interphase chromosomes to bind and be anchored. A large study, in which the human genome was probed for lamin B-binding sites, revealed 1,300 lamin-associated domains (LADs); many of these LADs were found to be in gene-poor regions of the genome (Guelen et al. 2008). Interactions with the nuclear lamina are associated with gene silencing and repression markers such as H3K4 dimethylation (Ferrai et al. 2010), increasing evidence for the idea that the periphery of the nucleus is associated with gene repression. Moreover, the disregulation of expression of both types of nuclear lamin has been correlated with cancer and degenerative disease (Butin-Israeli et al. 2012), including neurological degeneration (Coffinier et al. 2011).

The movement of chromatin in the nuclei appears to be largely constrained and thus reflects the physical attachment of chromatin to nuclear compartments, such as the nucleolus, nuclear periphery, and nucleoskeleton. Individual chromosomes occupy discrete compartments, and therefore distinct genomic regions localize to specific subnuclear positions. From several studies, it is becoming evident that nuclear position may have a crucial role for gene regulation. Moreover, it has been shown that there is a strong correlation between transcriptionally silent, late-replicating chromatin and a nuclear peripheral localization in several model systems (Boyle et al. 2001; Andrulis et al. 1998). Fluorescence recovery after photobleaching (FRAP) studies on mammalian cell nuclei indicate that in time periods of more than 1 h, chromatin becomes immobile over distances greater than 0.4 μm. Chubb et al. demonstrated that nucleoli and the nuclear envelope constrain the motion of interphase chromosomes that are located at these nuclear structures. In addition they demonstrated that the mobility of chromatin not associated with nucleoli or the nuclear periphery was much less constrained (Chubb et al. 2002).

Interphase Chromosome Positioning

We have been discussing how gene-poor regions of the genome are associated with the nuclear periphery. These data come from sophisticated global screening experiments. This distribution of more inactive areas of the genome at the nuclear periphery fits with earlier studies whereby whole gene-poor chromosomes were found to be located at the nuclear periphery. The first of these studies was performed by Bickmore and colleagues and demonstrated the differential distribution in interphase nuclei of the similarly sized chromosomes 18 and 19 (Croft et al. 1999). Chromosome 18, a gene-poor chromosome, was located at the nuclear periphery whereas chromosome 19, a gene-rich chromosome, was located in the nuclear interior. This gene density-correlated chromosome positioning in interphase nuclei was confirmed for all chromosomes in proliferating lymphoblastoid cells (Boyle et al. 2001). For human fibroblasts this gene density distribution is found in proliferating cells and not in nonproliferating cells (Bridger et al. 2000; Meaburn et al. 2007a, 2008; Mehta et al. 2007, 2010). Nonproliferating cells display a size-correlated distribution with large chromosomes toward the nuclear periphery and smaller chromosomes in the nuclear interior. Thus, when doing chromosome positioning studies it is critical to know whether the cells are proliferating. This point is especially important when comparing transformed and immortalized cancer cells with primary control cells, which will have a greater proportion of nonproliferating cells in the culture or tissue section. We use immune detection of the proliferation marker Ki-67, commonly used in neoplastic diagnostics (Kill 1996). The nuclei with very bright staining are in the proliferative cell cycle, and negative nuclei or nuclei with very dull staining are nonproliferating and are either quiescent or senescent. It is important that the cells with very dull staining are not counted as positive because this will lead to misinformation about proliferative status. In a primary culture of fibroblasts the maximum number of proliferating cells is usually never more than 65 %, and this is for the youngest of cultures. Therefore, pKi67 is a very important marker to use in chromosome positioning assays, but it must be analyzed correctly. During the past decade there have been a number of studies that have compared chromosome territory position between cancer cells and suitable control cells. However, very few of these have taken into account proliferative status.

We have found that individual chromosome territories change location in the cell nucleus when primary fibroblasts change proliferative status (Meaburn et al. 2007a; Mehta et al. 2007, 2010), meaning that some specific chromosomes are relocated whereas some stay where they are. Indeed, when cells are induced to become quiescent by serum starvation, interphase chromosomes either remain where they are, such as chromosome X, or move from a peripheral location to a more interior location, such as chromosomes 13 and 18, or move toward the nuclear periphery, as, for example, chromosome 10. We have shown that energy is required for the movement of these chromosomes, and nuclear motor proteins actin and nuclear myosin Iβ are involved. Others have also found that nuclear motor proteins are involved in chromatin relocation in the nucleus (Chuang et al. 2006; Dundr et al. 2007; Ondrej et al.

2007, 2008a, b). The idea that chromatin and chromosomes are translocated around the nucleus by nuclear motor proteins is a relatively new area of study for nuclear biologists, and as yet very little is known about the distribution and mode of action of the nuclear motor proteins themselves. Our studies have shown that NM1β is found throughout the nucleoplasm, with a concentration around the nuclear envelope and nucleoli in proliferating cells (Bridger and Mehta 2011; Mehta et al. 2010). It is extremely likely, given the importance and reproducibility of chromosome and gene positioning, that motor proteins involved in repositioning chromosomes and chromosomal subregions could be altered in disease states and cause issues for gene regulation. Indeed, we have observed in nonproliferating cells that NM1β distribution is very different, with large aggregates of the protein deep within the nucleoplasm. A similar distribution is apparent in cells derived from patients with the premature aging disease called Hutchinson–Gilford progeria syndrome (HGPS) (Mehta et al. 2011). Little or no research has been performed assessing nuclear myosins in cancer; however, one study correlated the presence of nuclear myosin VI with prognosis in renal cancer (Ronkainen et al. 2010) and another with nuclear myosin 18b in ovarian cancer (Yanaihara et al. 2004).

The Link Between Interphase Chromosome Location and Gene Expression

There is now evidence supporting the hypothesis that nuclear location of a chromosome and/or gene could play a role in regulating specific gene expression. For example, when resting human lymphocytes are activated by phytohemagglutinin, changes result in the intraorganization of chromosome territories, both in the degree of intermingling between territories and in their volume. More importantly, however, the radial positioning of the chromosome territories is changed. This alteration has been postulated to be a response to an altered transcriptional program (Branco et al. 2008). Furthermore, during ex vivo stem cell differentiation into adipocytes, the radial position of important genes involved in adipogenesis altered dramatically, with genes that become switched on when moving from the nuclear periphery toward the nuclear interior and back again when switched off. Control genes in this system that were either on or off did not respond to the adipogenic growth factors and did not change location (Szczerbal et al. 2009). In this differentiation system, there was little whole chromosome movement, but genes were looped out from chromosomes into the nuclear interior to associate with the nuclear structure SC35 speckles (Szczerbal and Bridger 2010).

Other studies have gone further, to identify where in the nucleus and to what nuclear structures the genes are targeted. Genes have been found to relocate to structures associated with active transcription and processing of RNA. Indeed, the activation of gene loci can involve a repositioning of genes toward areas of the nucleus where RNA polymerase II molecules aggregate into superstructures called transcription factories (Osborne et al. 2007). Other studies have shown genes becoming associated with other structures. For example, Dundr et al. inserted an artificial U2 snRNA array into the genome of cells and demonstrated that the array moved toward a stably positioned Cajal body for transcription through long-range chromosomal relocation. This movement was inhibited by an actin inhibitor, implying the involvement of actin in interphase chromosome repositioning (Dundr et al. 2007). Other studies have shown genes increasingly associated with SC35 domains upon upregulation (Brown et al. 2008; Szczerbal and Bridger 2010). It has also been shown that repositioning of genes from the periphery to a more interior position can correlate with inappropriate activation of that gene. The formation of chromatin loops for expression from repressed chromatin territories has been suggested as a mechanism of genome regulation, for example, for *Hox* gene activation (Chambeyron and Bickmore 2004). Indeed, most excitingly recent 3C (Ferraiuolo et al. 2010) and 4C conformation capture experiments have shown that actively transcribing Hox genes in a cluster are associated with a nuclear compartment for active transcription and that the nontranscribing genes are all located at a region where gene silencing occurs (Noordermeer et al. 2011). When the silenced genes become activated, they then co-compartmentalize with the other active *Hox* genes. This strict co-compartmentalization of genes explains the strict co-linearity rules associated with the *Hox* gene clusters where position in the cluster is correlated with the expression zone down the developing embryo.

The three-dimensional (3D) structure of the chromosomes within the territories also plays a major role in the control of gene expression. Regions of the chromosomes have been shown to interact with other regions of the same chromosome, in *cis*. For example, the locus control region (LCR) of the β-globin gene cluster acts as an enhancer of the β-globin genes, although it is more than 50 kb away. However, the LCR has been shown to be in close physical proximity to an actively transcribed HBB gene, suggesting a direct regulatory interaction (Carter et al. 2002; Tolhuis et al. 2002). This looping in 3D forms an active chromatin hub (ACH) for control of the expression of the β-globin genes (de Laat and Grosveld 2003), which dynamically associate with the LCR (Gribnau et al. 1998). As T-helper cells differentiate from naïve, uncommitted CD4-positive T cells, they show a transcriptional switch. Initially, the cells transcribe low levels of both Th1- and Th2-specific loci and regulators, but as they develop they become committed to either the Th1 or Th2 program (for review, see Murphy and Reiner 2002). Once a lineage has been established, it is retained as a heritable trait. This process of lineage commitment and differentiation involves the physical repositioning of regulators of gene expression. For example, it has been shown that during Th1 differentiation, the *GATA-3* and *c-maf* loci, which encode upstream regulators of Th2 cytokines, were progressively repositioned to centromeric heterochromatin and/or the nuclear periphery and repressed (Hewitt et al. 2004). These findings demonstrate another level of interphase chromosome behavior on gene expression, that is, that the intraorganization of a chromosome territory is also important (Fig. 2.2).

Noncoding RNAs (ncRNA) can control gene expression by establishing local repressive regions. For example, the *Air* ncRNA sets up a local 'cloud' of RNA that accumulates at the promoter of the imprinted *Slc22a3* gene and silences it by

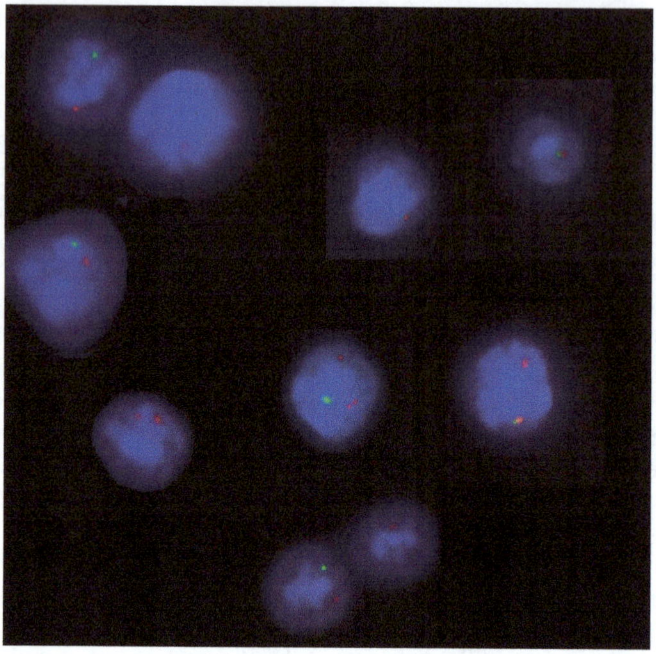

Fig. 2.2 Active genes can coassociate within the interphase nucleus. A montage panel of RNA FISH experiments demonstrates that the *Hbb-b1* gene loci (*green*) and the *Hba* gene loci (*red*) localize together more than 20 % of the time when actively transcribing. This interaction, shown in embryonic day 14.5 mouse erythroblasts, occurs although they are located on different chromosomes and demonstrates the preferential interaction of coregulated genes within the nuclear volume

recruiting G9a, an H3K9 histone methyltransferase (Nagano et al. 2008). The imprinted *Kcnq1* locus is also regulated by a paternally expressed repressive ncRNA, *Kcnq1ot1*, which regulates a domain of up to 750 kb. However, local activation of genes may be able to overcome the regional silencing effects of ncRNAs, as Kcnq1 transitions from monoallelic to biallelic expression during the development of the heart, and there have recently been shown to be both tissue- and stage-specific chromatin loops between the Kcnq1 promoter and newly identified DNA regulatory elements (Korostowski et al. 2011). The most notable example of ncRNAs silencing genes is X-chromosome inactivation, where the ncRNA *Xist* silences an entire chromosome. The *Xist* ncRNA covers the chromosome that is going to inactivate and condenses into a smaller, compact structure, which is associated with the periphery (Clemson et al. 1996). As silencing is established, a repressive nuclear compartment forms that excludes RNA polymerase II and transcription factors. Transcriptional repression follows the formation of this compartment, possibly as genes become physically pulled down into the repressive environment, where they are inaccessible to the transcriptional machinery (Chaumeil et al. 2006). The few genes that remain expressed, for example, those in the pseudoautosomal region, loop out of the repressive compartment to be expressed (Splinter et al. 2011).

Interphase Chromosome Behavior in Hutchinson–Gilford Progeria Syndrome

Hutchinson–Gilford progeria syndrome (HGPS) is a severe premature aging disease that affects children. First described by Jonathan Hutchinson and Hastings Gilford in the 1800s, this disease is recognized by a group of characteristics indicative of premature aging. The most common of these include alopecia (hair loss), failure to thrive (short stature and low weight), lipodystrophy (loss of fat), scleroderma of the skin, and increased visibility of blood vessels. Initially, children appear unaffected, but symptoms usually present around 1 year of age, leading to a mean age of diagnosis of 2.9 years. Another main characteristic of HGPS is heart disease. Patients suffer from atherosclerosis/hardening of the arteries, which is sometimes associated with calcification. This change, in combination with loss of smooth muscle from blood vessels, leads to an increased risk of heart attacks and stroke. These are the main causes of death in this disease, with the average life expectancy of a HGPS patient being 13.5 years (Hennekam 2006).

HGPS is an extremely rare disease with an incidence of approximately 1 in every 4–8 million live births. Of these cases approximately 80 % are caused by the same de novo mutation in the *LMNA* gene (De Sandre-Giovannoli et al. 2003; Eriksson et al. 2003). This gene encodes both A-type lamins, making HGPS part of a group of diseases known as the laminopathies. The 'classic' mutation found in the majority of HGPS patients is the G608G mutation, which is a silent mutation at the protein level. At the DNA level, however, it causes activation of a cryptic splice donor site, which results in an interstitial deletion of 150 amino acids from exon 11. This deletion gives rise to a truncated protein, with a 50-amino-acid deletion, called Progerin (Eriksson et al. 2003). Pre-lamin A and progerin are subject to the same posttranslational modifications. The region deleted in progerin contained an important cleavage site for the enzyme ZMPSTE24, which removes the farnesylated N-terminus of the protein, freeing it from the membrane. Lacking the cleavage site, progerin therefore remains bound to the nuclear membrane. Interestingly, homozygous mutations in the ZMPSTE24 gene have also been found to cause an atypical form of HGPS. Progerin expression is thought to have a dominant negative effect on cell function; it has been shown to cause thickening of the nuclear envelope as well as nuclear shape abnormalities such as blebs and invaginations (Goldman et al. 2004; Bridger and Kill 2004).

The association of the nuclear lamina with the chromatin and chromosomes increased interest in genome organization and chromosome/chromatin localization in cases of disruption to the nuclear lamina through mutation in *LMNA*, such as that seen in HGPS. Genome organization has been shown to be disrupted in a number of cells with lamin A mutation or that lack lamin A completely (Galiova et al. 2008; Shimi et al. 2008: Taimen et al. 2009). Further, three studies have revealed mispositioning of whole chromosome territories in cells with *LMNA* mutations (Meaburn et al. 2007a; Mewborn et al. 2010; Mehta et al. 2011). Interestingly, using chromosome 10 positioning that had previously been shown to occupy different nuclear locations in proliferating, quiescent, and senescent cells (Mehta et al. 2007, 2010), Mehta et al. (2011) revealed that HGPS cells had a quiescent-type distribution of this chromosome in proliferating HGPS fibroblasts. Complete reorganization of the genome was, however, not observed because the X chromosomes were found at the nuclear periphery in both control and HGPS cells. This mislocalization of chromosomes could be restored to normal when the HGPS cells were treated with farnesyl transferase inhibitors that prevent progerin from being farnesylated. A global genome-wide study of the sequences associated with progerin at the nuclear periphery in mouse cells confirms that A-type lamins are involved in chromatin and genome organization in nuclei. Kubben et al. (2012) show that in cells with progerin some genes have been relocated away from the nuclear periphery, whereas others have enhanced association.

Cancer

Cancer is a disease characterised by genomic instability, resulting in unlimited cell replicative potential. Transformation is a multistep process usually encompassing many genetic modifications including aneuploidy, copy number variants, gene mutations, aberrant DNA methylation patterns, and chromosomal rearrangements. The majority of these changes promote increased oncogenic transcription, which stimulates proliferation and inhibits apoptosis.

With increased understanding of chromosome territories in the interphase nucleus, much work has gone into understanding the differences that emerge in neoplastic tissue compared with normal samples. The observed changes have been both on the global scale, such as loss of heterochromatin, and at the gene scale, such as the repositioning of tumor-associated genes in cancer formation (Zhu et al. 2011; Meaburn and Misteli 2008).

The changes in the nuclear architecture of cancer cells are so robust they have been used in tumor diagnosis for more than 140 years. Since the first patient biopsy was examined in 1860, many advances have been made in understanding cancer. However, diagnosis still relies heavily on the analysis of cell morphology. Specific nuclear markers of cancer include changes in nuclear size and shape, nucleolus alterations, changes in chromatin organization, aberrantly shaped nuclear lamina, and alterations to promyelocytic leukemia (PML) bodies (Zink et al. 2004). Common nuclear shape changes include indentations and folds that are indicative of a wide variety of cancers. Nuclear morphological changes with specificity to certain cancer types include grooves or long clefts in the nuclear surface, which are associated with the expression of the papillary thyroid oncogene expressed exclusively in papillary thyroid carcinomas (Fischer et al. 1998). Enlarged nucleoli are associated with several cancer types; however, inconspicuous nucleoli are almost exclusively indicative of small-cell anaplastic lung carcinoma (Zink et al. 2004). Observed changes to chromatin structures include changes to heterochromatic foci, which are areas of the nucleus that contain highly compact chromatin structures usually associated with gene silencing (Hahn et al. 2010). The changes to heterochromatic foci include loss, asymmetry, coarse appearance, and spreading throughout the nucleus (Zink et al. 2004). Several studies have identified silencing of tumor suppressor genes in cancer in parallel with changes to chromatin structure (Hahn et al. 2010). Tumor suppressor gene promoters showing heterochromatic markers such as H3K9 trimethylation have been identified in many cancer types (Lakshmikuttyamma et al. 2010). It is yet to be established whether changes in chromatin structure cause silencing of tumor suppressor genes and thus drive cancer. One study conversely found that knockdown mice lacking the tumor suppressor gene *BRCA1* resulted in changes to heterochromatin, including loss of foci from the nuclear periphery, leading to a more diffuse state of foci throughout the nucleus (Zhu et al. 2011). This finding suggests some sort of positive feedback whereby tumor suppressors regulate chromatin conformation; but once lost, aberrant chromatin changes promote cancer.

Chromosome Positioning in Cancer

A vast number of diseases present with genetic defects that are often visible as chromosome rearrangements. The presence of chromosome abnormalities is a hallmark for many forms of cancer. In many cases the specific association of certain chromosome aberrations and type of tumor are considered of diagnostic and prognostic value.

Studies on interphase chromosome position in cancer cells were initiated by work where HT1080 fibrosarcoma cell transformation was chemically reversed. Acrocentric chromosomes that had been found through the nucleus were relocated more centrally, which seems to result from their association with the nucleoli and the coalescence of many smaller nucleoli to one prominent centrally located nucleolus after the treatment (Krystosek 1998). In a study from the Cremer laboratory, the differential positioning of human chromosomes 18 and 19 was much less obvious in colon adenocarcinoma cells, cervix carcinoma cells, and Hodgkin disease-derived cells (Cremer et al. 2003). The nuclear locations of chromosomes 10, 18, and 19 were assessed in normal thyroid tissue and compared to adenomatous goiters, papillary carcinomas, and undifferentiated carcinomas. There was no difference in chromosome position in the normal and goiter tissue with chromosomes 10 and 18 positioned toward the nuclear periphery; and chromosome 19 in a central location. However, in the papillary carcinoma tissue chromosome 19 was located centrally in statistically fewer cells. Further, in undifferentiated carcinomas all the chromosomes assessed were mislocalized (Murata et al. 2007). In a breast cancer cell line MCFCA1a differences in the distance between chromosomes 4 and 16 were found when compared to the control cell line MCF10A (Marella et al. 2009). Wiech et al. (2005) analyzed chromosome 8 positions in wax-embedded pancreatic cancer tissue samples. Radial distance indicated the repositioning of chromosome 8 to the nuclear periphery, which matched roundness scores showing a change in the shape of the territory. A subsequent paper also noted a reduction of the roundness of chromosome 8 territories, suggesting a thinner, more elongated territory in pancreatic carcinomas (Timme et al. 2011). The centromere and the gene encoding HER2 on chromosome 17 were also shown to compact in neoplastic breast tissue, conferring the repositioning of the centromere to a more internal location (Wiech et al. 2005). A subsequent study by Wiech et al. (2009) reported repositioning of chromosome 18 during cell differentiation of nonneoplastic cervical squamous epithelium, showing a move toward the nuclear interior. This finding was in contrast to the observations in cervical squamous carcinomas that showed a repositioning of chromosome 18 toward the nuclear periphery (Wiech et al. 2009). This study also analyzed the expression levels of *BCL2*, an inhibitor of apoptosis, which was shown to prolong cell survival and found to be unregulated in 54 % of cervical cancers. A reduction in *BCL2* expression has been found in the terminally differentiated cells on the outer layers of the cervical epithelium. In contrast, an increase in *BCL2* expression was found in the carcinomas, suggesting that relocation to the nuclear periphery increases *BCL2* transcription (Wiech et al. 2009).

Gene Repositioning in Interphase of Cancer Cells

A comprehensive study into the nuclear organization in breast cancer by Meaburn et al. (2009) found that 8 of the 20 gene loci analyzed showed significant gene repositioning in cancer cells. All the gene loci studied have previously been implicated in cancer, with the most frequently repositioned gene locus being *HES5*, a transcription repressor that regulates cell differentiation. As the majority of genes were not repositioned, this finding suggests that the repositioning was gene specific rather than global genome reorganization. It was also concluded that gene repositioning was not associated with genome ploidy because the genes analyzed in this study had no changes in copy number. It was also observed that some genes were only repositioned in certain cancer types, suggesting that some gene repositioning is cancer type specific (Meaburn et al. 2009). This idea is supported by the findings of Wiech et al. (2005), who identified *BCL2* repositioning to the periphery in *BCL2*-positive cervical squamous cell carcinomas but not in *BCL2*-negative cancer cells. By analyzing gene position in normal cells and cells from noncancerous disease, breast hyperplasia or fibroadenoma, no significant difference was found (Meaburn et al. 2009). This result demonstrated that the rearrangements observed in cancer cells are cancer specific and cannot be seen in noncancerous diseased cells. The identification of cancer-specific genes repositioned in several types of breast cancer could prove a useful diagnostic tool. One problem this technique faces, however, is the intermingling of normal and diseased cells, which reduces the statistical power. It has been observed that the tissue directly adjacent to the cancerous tissue in patients has a normal pattern of gene organization that matched with the tissue from normal individuals (Meaburn et al. 2009). This finding is in agreement with previous reports that the organization of genes does not differ between individuals, except in the case of disease (Wiech et al. 2005). Once validated in a larger number of patient samples, these problems should be overcome, yielding a useful diagnostic tool. Although the study by Meaburn et al. (2009) failed to show a relationship between gene repositioning and transcription levels, other studies investigating this aspect have shown that an altered positioning of specific genes in the nucleus is associated with altered transcription levels (Wiech et al. 2009). This finding is in contrast with a previous report that showed that gene repositioning occurs in early tumorigenesis and does not affect transcription levels (Meaburn and Misteli 2008). More work is needed in this area to understand why gene repositioning occurs, especially if it is not related to gene function. Seminal studies aimed at understanding differentiation and maturation of the lymphoid lineage have analyzed the relationship between gene positioning and activity. One of these studies showed preferential localization to the nuclear interior of the IGH and IGk loci during pro-B-cell lymphocyte development (Kosak et al. 2002). This repositioning event coincided with transcription, and subsequent recombination of these loci is required for the production of unique antibodies. This finding led the authors to conclude that chromatin rearrangement is a powerful mechanism for the control of transcription (Kosak et al. 2002). Further studies will elucidate the role of gene positioning within the nuclear architecture as an underlying condition for gene transcription and expression in cancer development and progression.

Formation of Chromosome Translocations in the Context of Nuclear Organization

Cancer cells harbor a number of genetic abnormalities, of which chromosomal translocations are well-studied examples, especially in leukemia and lymphoma. The mechanisms of translocation formation are under study. Whether a multistep process or a simultaneous occurrence of several events (Forment et al. 2012), we expect an impact on genome organization and nuclear architecture. In any case, the exchange of chromosomal fragments requires the formation of two or more double-strand breaks (DSB). The incorrect repair of DSB leads to the fusion of nonhomologous chromosome ends, creating derivative chromosomes. The most error prone pathway for the repair of DSB is nonhomologous end-joining, in which two chromosome ends in close proximity are religated. Within the nucleus there are error-free repair pathways for DSB resolution, such as homologous recombination; however, this process does require either a sister chromatid (post S-phase) or homologous chromosome (Meaburn et al. 2007b). There is an ever-growing list of cancer type-specific translocations, with the same rearrangement arising nonrandomly, and hence observed in the cancer cells of many individuals (Mitelman Database of Chromosome Aberrations in Cancer, http://cgap.nci.nih.gov/Chromosomes/Mitelman). These recurrent translocations are useful diagnostic tools and are often associated with clinical outcome. A classical example is the Philadelphia chromosome derived from the t(9;22) and found in chronic myeloid leukemia (CML) and in some cases of acute lymphoblastic leukemia (ALL) (Goldman 2010). Another example is the t(12;21), found in approximately one third of pediatric patients with ALL and associated with a relatively good prognosis (Harrison et al. 2010). With increased understanding of the organization of the interphase nucleus, a number of studies have investigated the proximity of chromosomes involved in common translocations as well as the 3D positioning of their derivatives (Meaburn et al. 2007b) in certain cancers. There is evidence to support the hypothesis that translocations occur in interphase nuclei between chromosomes that occupy similar nuclear space (Kozubek et al. 1999; Parada et al. 2002; Kuroda et al. 2004; Gandhi et al. 2012); this may also be true for intrachromosomal fusions and genes that are at some distance linearly but may be placed together by chromosome folding (Gandhi et al. 2006). Nuclear position was also identified as a factor that contributed to translocation frequency, with peripherally located chromosomes such as 4, 13, and 18 being involved in a higher than expected number of translocations (Bickmore and Teague 2002). Later global screening studies have confirmed that nuclear position is fundamental in the selection of translocation partners (Engreitz et al. 2012; Roix et al. 2003), but transcriptional activity is also of fundamental importance (Klein et al. 2011). A number of studies have been undertaken to understand the interaction of different chromosome regions that favor the exchange of DNA fragments at the level of interphase nucleus, hence the formation of chromosome translocations (Branco and Pombo 2006; Murmann et al. 2005; Roix et al. 2003; Zhang et al. 2012). This point is supported by the finding that chromosome territories do not have neat borders and that neighboring territories do intermingle (Branco and Pombo 2006). It has been suggested that chromosomal translocations are events whose frequency is correlated to the spatial proximity of the loci involved, as described for some human lymphomas (Roix et al. 2003). One could speculate that a similar location in the nuclear environment is sufficient to facilitate an encounter and an exchange of chromosome fragments, a phenomenon described as chromosome kissing (Cavalli 2007). Also, gene loci located on the periphery of chromosome territories were found to be involved in more interchromosomal rearrangements than those deep within the territory (Gandhi et al. 2009). Internally located loci were more frequently involved in intrachromosomal aberrations (Gandhi et al. 2009).

Exposure to ionizing radiation results in DNA DSBs that permit nonhomologous chromosomes in close proximity to combine, creating complex rearrangements (Anderson et al. 2002). This rearrangement can cause several cancer types, in particular, radiation-induced thyroid tumors, such as papillary thyroid cancer. A common chromosome rearrangement observed in papillary thyroid cancer is the intrachromosomal inversion on chromosome 10 that creates the fusion gene *RET/PTC1*. Although the two genes involved in this inversion are on the same chromosome, they are 30 Mb apart. One study, however, showed that at least one copy of each gene colocalized in 35 % of normal thyroid tissues compared with only 6 %

in mammary epithelial tissue (Nikiforova et al. 2000); this explains the tendency to form inversions specifically in the thyroid. Another translocation that may arise from exposure to ionizing radiation is the t(9;22)(q34;q11). This translocation gives rise to the Philadelphia chromosome and the *BCR-ABL* fusion gene. The oncogenic chimeric protein produced drives the formation of CML as well as some cases of ALL. It has been shown that the *BCR* and *ABL* genes are found in distinct locations in the interphase nucleus of healthy stimulated and nonstimulated lymphocytes (Lukasova et al. 1997). In response to ionizing radiation, however, both genes were shown to move to a more internal location, reducing the distance between them to less than 1 μm in 47.5 % of healthy donors (Lukasova et al. 1997; Kozubek et al. 1997). This finding suggests that rearrangement of chromatin in response to ionizing radiation brings into close proximity two genes known to be common translocation partners.

Regions from different chromosomes can also be brought into close proximity by association with specific nuclear structures such as nucleoli (Sullivan et al. 2001). The acrocentric chromosomes (13, 14, 15, 21, and 22) of the human genome all contain nucleolar organizer regions (NOR), which are composed of ribosomal gene repeats. Nucleoli form around these NOR elements after mitosis, and during cell-cycle progression the nucleoli fuse, creating fewer larger structures. However, it should be noted that this process is very rarely observed in cancer (Morgan et al. 1987), whereas it is more likely to affect gamete formation and offspring as a result of loss of genetic material. This process does, however, demonstrate that association with nuclear elements can increase the occurrence of translocations. One example specific to cancer is the translocation observed between the mouse chromosomes 12 and 15, which is present in 80 % of plasmacytomas (Osborne et al. 2007). The breakpoints in this translocation involve the *c-Myc* gene and immunoglobulin heavy chain locus (*IgH*). This translocation is mirrored in humans by that of t(8;14), which encompasses the same genes and is found in Burkitt's lymphoma as well as other forms of lymphoid cancers (Haluska et al. 1987). These genes are found in close proximity in only a third of human nuclei but are neighbors in mouse cells (Parada et al. 2004; Roix et al. 2003). A study by Osborne et al. (2007) found that upon activation both genes are recruited to the same transcription factories, increasing their physical proximity. As proximity has been shown to be a key factor in translocations, this increases our understanding of why the t(12;15) is observed so frequently. In support of this realization, it was observed that *c-Myc* colocalized and transcribed with *IgH* at the expected frequency to give rise to the observed level of translocations (Osborne et al. 2007). The other translocation partners of *c-Myc* in Burkitt's lymphoma and plasmocytoma include *IgK* and *IgL*. The colocalization of *c-Myc* to these genes was also analyzed and found to correlate with translocation frequencies. Therefore, this research suggests a correlation between the number of times genes come into close proximity and the likelihood of translocations (Osborne et al. 2007).

Proximity is only one factor thought to affect translocation frequencies; other factors to consider include chromosome size and gene density (Bickmore and

Teague 2002). One study found a correlation between chromosome size and translocation frequency in response to ionizing radiation (Cafourkova et al. 2001). Another comprehensive study that analyzed more than 11,000 non-disease-causing chromosome aberrations found that larger chromosomes were more frequently involved in translocations; this could, however, be ascribed to increased opportunity for translocations in large chromosomes because they are bigger targets. This study also identified that translocations appear to occur less frequently in highly dense regions of the genome (Bickmore and Teague 2002).

Repositioning of Genes Affected by Translocation Events

It has been postulated that, because of a translocation event, specific genes might alter their position in the nucleus and therefore be more or less exposed to the transcription machinery. More precisely, certain genes could be activated or inactivated on the basis of the new environment they inhabit. This change would happen when two different regions characterized by different transcriptional activity become positioned next to one another (as in the case of a reciprocal translocation), resulting in an aberrant localization in the nucleus for one or both of the two regions. Studies on both constitutional syndromes and cancer have focused on the localization of the derivative chromosomes in the cell nucleus and also explored gene expression in the context of the newly established nuclear architecture (Ballabio et al. 2009; Harewood et al. 2010). A study on Ewing sarcoma cells has shown that the fusion genes derived from the cancer-associated rearrangement t(11;22) assume an intermediate nuclear position when compared to the wild type *EWSR1* and *FLI1* genes (Taslerova et al. 2003). Murmann and coworkers observed that the change in position of loci affected by a translocation depends on the relative gene density of the 2-Mb window of the region considered. The study of wild-type *MLL* and five of its translocation partners showed that the resulting fusion genes changed their nuclear location according to the reciprocal gene density of the region involved (Murmann et al. 2005). More recently, a study on pediatric leukemia characterized by the presence of the acquired t(7;12) translocation has shown that an overexpression of the *HLXB9* gene (on chromosome 7q36) corresponded to an altered nuclear position of the derivative chromosome carrying the *HLXB9* gene itself (Ballabio et al. 2009). In this case, the translocated *HLXB9* gene localized more centrally than the wild-type allele. A larger study on the constitutional balanced translocation t(11;22) has shown on a larger scale that an altered spatial organization of the der(11) corresponds to an alteration of the expression profile of genes localized on the der(11). In the same study, chromosomes other than those involved in the rearrangement have also shown an altered nuclear position and altered gene expression profiles (Harewood et al. 2010). This finding shows that the global nuclear architecture and the location of various chromosomes are influenced by specific rearrangements. Altogether, nuclear positioning plays a functional role in regulating gene expression.

Aneuploidy and DNA Copy Number Alterations

Gain and loss of genetic material is another common feature of cancer cells. Microscopically, this defect can be visualized as complete loss or gain of entire chromosomes caused by missegregation during mitosis and resulting in aneuploidy. Other imbalances are visible as loss or gain of certain chromosomal regions: these are known as deletions, duplications, amplifications, or more generally as DNA copy number alterations (CNA). These changes have an impact in diagnosis and are relevant at the prognostic level. For example, extra copies of chromosome 3q define the difference between cervical dysplasia and invasive cervical carcinoma resulting from human papilloma virus (HPV) infection, whereas complete or partial loss of chromosomes 5 and 7 are the most commonly observed alterations in acute myeloid leukaemia (AML) (Zhang et al. 2011). It is assumed that the presence of additional genetic material corresponds to increased expression levels of the overrepresented sequences. This assumption is supported by a study on highly hyperdiploid pediatric ALL, showing that the presence of additional chromosomal material corresponded to an increased expression of the amplified loci (Gruszka-Westwood et al. 2004).

Very few studies have addressed the issue of chromosome organization in cases of aneuploidy. Croft et al. (1999) did not see any repositioning of an extra chromosome 18 in Edward syndrome cells with a trisomy 18 (Croft et al. 1999). This finding is supported by that of Koutna et al. (2000), who investigated specific trisomic loci within the HT-29 colon cancer cell line. They concluded that the location of the third copy of a specific locus is not significantly relocated when compared to the two loci present in a noncancerous tissue (Koutna et al. 2000). Although gene amplification resulting from aneuploidy has the ability to drive cancer formation it does not alter the organization of chromosome territories. Therefore, according to these studies, tumorigenesis appears to be independent of chromosome position. In another study, additional copies of chromosomes 7, 18, or 19 were artificially introduced in immortalized or cancer cell lines, and their position in the nucleus was observed and correlated with altered expression levels. It was ascertained that the presence of additional chromosomes increased transcription from the trisomic loci. However, a shift in positioning was noted for chromosomes 18 and 19, but not for chromosome 7. The authors proposed that positioning within the nucleus is determined by a unique chromosome-specific 'zip code' that might be independent from the transcriptional activity of the sequences that compose it (Sengupta et al. 2007).

Other Disease Situations

In 1988, Manuelidis and Borden published their seminal work demonstrating that specific chromosomal domains were located to specific regions of the nuclei of neurons and glial cells. In large neurons, probes delineating chromosomes 9, 1, and Y were most commonly found adjacent to nucleoli. However, in astrocytes these same regions were found at the nuclear membrane and not specifically associated with nucleoli. These data indicate that nonrandom chromosome positioning is of importance to the cell even in terminally differentiated cells such as nerve cells. Manuelidis was the first to show in the human cortex the spatial repositioning of chromosome in interphase nuclei in disease. She found that chromosome X had become relocated from the nuclear edge to the nuclear interior in seizure foci in epileptic patients. This study is significant because it links chromosome positioning with ill health. In this study we do not know if the repositioning affects gene expression on the X chromosome. However, one of the master regulator genes for epilepsy has been identified on the X chromosome (Stromme et al. 2002).

Very few studies have concerned chromosome and gene repositioning after an infection. However, genes have been observed to relocate within cells of hosts that are exposed to infectious agents. In *Biomphalaria glabrata* cells, the secondary host organism of the human parasitic disease schistosomiasis, commonly known as bilharzia, specific genes involved in the infection become relocated within the interphase nuclei at the same time that quantitative polymerase chain reaction (PCR) reveals that they are being expressed (Knight et al. 2011; Arican, Ittisprasert, Bridger, and Knight, manuscript in preparation). One other study revealed chromosome 17 and not 18 changed nuclear location over time after an Epstein–Barr virus (EBV) infection (Li et al. 2010).

Concluding Remarks

As more laboratories consider the 3D and 4D nuclear organization of the genome in their studies on genome function, it is becoming clearer that chromosome position and association with nuclear structure matter a great deal with respect to regulating gene expression in healthy cells and affect the functioning of diseased cells when misorganization of the chromosomes and genes is apparent. Furthermore, misorganization and misplacement of chromosomes and their gene loci may be responsible for some disease situations.

References

Anderson RM, Stevens DL, Goodhead DT (2002) M-FISH analysis shows that complex chromosome aberrations induced by alpha-particle tracks are cumulative products of localized rearrangements. Proc Natl Acad Sci U S A 99:12167–12172

Andrulis ED, Neiman AM, Zappulla DC, Sternglanz R (1998) Perinuclear localization of chromatin facilitates transcriptional silencing. Nature (Lond) 394:592–595

Ballabio E, Cantarella CD, Federico C, Di Mare P, Hall G, Harbott J, Hughes J, Saccone S, Tosi S (2009) Ectopic expression of the HLXB9 gene is associated with an altered nuclear position in T(7;12) leukaemias. Leukemia 23:1179–1182

Bickmore WA, Teague P (2002) Influences of chromosome size, gene density and nuclear position on the frequency of constitutional translocations in the human population. Chromosome Res 10:707–715

Boisvert FM, Van Koningsbruggen S, Navacues J, Lamond AI (2007) The multifunctional nucleolus. Nat Rev Mol Cell Biol 8:574–585

Boyle S, Gilchrist S, Bridger JM, Mahy NL, Ellis JA, Bickmore WA (2001) The spatial organization of human chromosomes within the nuclei of normal and emerin-mutant cells. Hum Mol Genet 10:211–219

Branco MR, Pombo A (2006) Intermingling of chromosome territories in interphase suggests role in translocations and transcription-dependent associations. PLoS Biol 4:e138

Branco MR, Branco T, Ramirez F, Pombo A (2008) Changes in chromosome organization during PHA-activation of resting human lymphocytes measured by cryo-FISH. Chromosome Res 16:413–426

Bridger JM, Kill IR (2004) Aging of Hutchinson-Gilford progeria syndrome fibroblasts is characterised by hyperproliferation and increased apoptosis. Exp Gerontol 39:717–724

Bridger JM, Kill IR, Lichter P (1998) Association of pKi-67 with satellite DNA of the human genome in early G_1 cells. Chromosome Res 6:13–24

Bridger JM, Boyle S, Kill IR, Bickmore WA (2000) Re-modelling of nuclear architecture in quiescent and senescent human fibroblasts. Curr Biol 10:149–152

Bridger JM, Mehta IS (2011) Nuclear molecular motors for active, directed chromatin movement in interphase nuclei. Advances in nuclear architecture. Eds Niall Adams and Paul Freemont. Springer.

Brown JM, Green J, das Neves RP, Wallace HA, Smith AJ, Hughes J, Gray N, Taylor S, Wood WG, Higgs DR, Iborra FJ, Buckle VJ (2008) Association between active genes occurs at nuclear speckles and is modulated by chromatin environment. J Cell Biol 182:1083–1097

Butin-Israeli V, Adam SA, Goldman AE, Goldman RD (2012) Nuclear lamin functions and disease. Trends Genet 28:464–471

Cafourkova A, Luka-ova E, Kozubek S, Kozubek M, Govorun RD, Koutna I, Bartova E, Skalnikova M, Jirsova P, Pasekova R, Krasavin EA (2001) Exchange aberrations among 11 chromosomes of human lymphocytes induced by gamma-rays. Int J Radiat Biol 77:419–429

Carter D, Chakalova L, Osborne CS, Dai Y-F, Fraser P (2002) Long-range chromatin regulatory interactions in vivo. Nat Genet 32(4):623–626. doi:10.1038/ng1051

Cavalli G (2007) Chromosome kissing. Curr Opin Genet Dev 17:443–450

Chambeyron S, Bickmore WA (2004) Chromatin decondensation and nuclear reorganization of the HoxB locus upon induction of transcription. Genes Dev 18:1119–1130

Chaumeil J, Baccon PL, Wutz A, Heard E (2006) A novel role for Xist RNA in the formation of a repressive nuclear compartment into which genes are recruited when silenced. Genes Dev 20:2223–2237

Chuang CH, Carpenter AE, Fuchsova B, Johnson T, de Lanerolle P, Belmont AS (2006) Long-range directional movement of an interphase chromosome site. Curr Biol 16:825–831

Chubb JR, Boyle S, Perry P, Bickmore WA (2002) Chromatin motion is constrained by association with nuclear compartments in human cells. Curr Biol 12(6)439–445.

Clemson CM, McNeil JA, Willard HF, Lawrence JB (1996) XIST RNA paints the inactive X chromosome at interphase: evidence for a novel RNA involved in nuclear/chromosome structure. J Cell Biol 132:259–275

Coffinier C, Jung HJ, Nobumori C, Chang S, Tu Y, Barnes RH II, Yoshinaga Y, de Jong PJ, Vergnes L, Reue K, Fong LG, Young SG (2011) Deficiencies in lamin B1 and lamin B2 cause neurodevelopmental defects and distinct nuclear shape abnormalities in neurons. Mol Biol Cell 22:4683–4693

Cremer M, Kupper K, Wagler B, Wizelman L, von Hase J, Weiland Y, Kreja L, Diebold J, Speicher MR, Cremer T (2003) Inheritance of gene density-related higher order chromatin arrangements in normal and tumor cell nuclei. J Cell Biol 162:809–820

Croft JA, Bridger JM, Boyle S, Perry P, Teague P, Bickmore WA (1999) Differences in the localization and morphology of chromosomes in the human nucleus. J Cell Biol 145:1119–1131
de Laat W, Grosveld F (2003) Spatial organization of gene expression: the active chromatin hub. Chromosome Res 11:447–459
De Sandre-Giovannoli A, Bernard R, Cau P, Navarro C, Amiel J, Boccaccio I, Lyonnet S, Stewart CL, Munnich A, Le Merrer M, Levy N (2003) Lamin A truncation in Hutchinson–Gilford progeria. Science 300:2055
Dundr M, Ospina JK, Sung MH, John S, Upender M, Ried T, Hager GL, Matera AG (2007) Actin-dependent intranuclear repositioning of an active gene locus in vivo. J Cell Biol 179:1095–1103
Engreitz JM, Agarwala V, Mirny LA (2012) Three-dimensional genome architecture influences partner selection for chromosomal translocations in human disease. PLoS One 7:e44196
Eriksson M, Brown WT, Gordon LB, Glynn MW, Singer J, Scott L, Erdos MR, Robbins CM, Moses TY, Berglund P, Dutra A, Pak E, Durkin S, Csoka AB, Boehnke M, Glover TW, Collins FS (2003) Recurrent de novo point mutation in lamin A cause Hutchinson–Gilford progeria syndrome. Nature (Lond) 423:293–298
Ferrai C, de Castro IJ, Lavitas L, Chotalia M, Pombo A (2010) Gene positioning. Cold Spring Harb Perspect Biol 2:a000588
Ferraiuolo MA, Rousseau M, Miyamoto C, Shenker S, Wang XQ, Nadler M, Blanchette M, Dostie J (2010) The three-dimensional architecture of Hox cluster silencing. Nucleic Acids Res 38:7472–7484
Feuerbach F, Galy V, Trelles-Sticken E, Fromont-Racine M, Jacquier A, Gilson E, Olivo-Marin JC, Scherthan H, Nehrbass U (2002) Nuclear architecture and spatial positioning help establish transcriptional states of telomeres in yeast. Nat Cell Biol 4:214–221
Fischer AH, Bond JA, Taysavang P, Battles OE, Wynford-Thomas D (1998) Papillary thyroid carcinoma oncogene (RET/PTC) alters the nuclear envelope and chromatin structure. Am J Pathol 153:1443–1450
Forment JV, Kaidi A, Jackson SP (2012) Chromothripsis and cancer: causes and consequences of chromosome shattering. Nat Rev Cancer 12:663–670
Foster HA, Bridger JM (2005) The genome and the nucleus: a marriage made by evolution. Genome organisation and nuclear architecture. Chromosoma (Berl) 114:212–229
Galiova G, Bartova E, Raska I, Krejci J, Kozubek S (2008) Chromatin changes induced by lamin A/C deficiency and the histone deacetylase inhibitor trichostatin A. Eur J Cell Biol 87:291–303
Gandhi M, Medvedovic M, Stringer JR, Nikiforov YE (2006) Interphase chromosome folding determines spatial proximity of genes participating in carcinogenic RET/PTC rearrangements. Oncogene 25:2360–2366
Gandhi MS, Stringer JR, Nikiforova MN, Medvedovic M, Nikiforov YE (2009) Gene position within chromosome territories correlates with their involvement in distinct rearrangement types in thyroid cancer cells. Genes Chromosomes Cancer 48:222–228
Gandhi M et al (2012) Frequency of close positioning of chromosomal loci detected by FRET correlates with their participation in carcinogenic rearrangements in human cells. Genes Chromosomes Cancer 51:1037–44
Goldman JM (2010) Chronic myeloid leukaemia: a historical perspective. Semin Hematol 47:302–311
Goldman RD, Shumaker DK, Erdos MR, Eriksson M, Goldman AE, Gordon LB, Gruenbaum Y, Khuon S, Mendez M, Varga R, Collins FS (2004) Accumulation of mutant lamin A causes progressive changes in nuclear architecture in Hutchinson–Gilford progeria syndrome. Proc Natl Acad Sci U S A 101:8963–8968
Gomez-Cavazos JS, Hetzer MW (2012) Outfits for different occasions: tissue-specific roles of nuclear envelope proteins. Curr Opin Cell Biol 24(6):775–783
Gribnau J, de Boer E, Trimborn T, Wijgerde M, Milot E, Grosveld F, Fraser P (1998) Chromatin interaction mechanism of transcriptional control in vivo. EMBO J 17:6020–6027

Gruszka-Westwood AM et al (2004) Comparative expressed sequence hybridization studies of high-hyperdiploid childhood acute lymphoblastic leukemia. Genes Chromosomes Cancer 41:191–202

Guelen L, Pagie L, Brasset E, Meuleman W, Faza MB, Talhout W, Eussen BH, de Klein A, Wessels L, de Laat W, van Steensel B (2008) Domain organization of human chromosomes revealed by mapping of nuclear lamina interactions. Nature (Lond) 453:948–951

Hahn M, Dambacher S, Schotta G (2010) Heterochromatin dysregulation in human diseases. J Appl Physiol 109:232–242

Haluska FG, Tsujimoto Y, Croce CM (1987) The t(8;14) chromosome translocation of the Burkitt lymphoma cell line Daudi occurred during immunoglobulin gene rearrangement and involved the heavy chain diversity region. Proc Natl Acad Sci U S A 84:6835–6839

Harewood L, Schutz F, Boyle S, Perry P, Delorenzi M, Bickmore WA, Reymond A (2010) The effect of translocation-induced nuclear reorganization on gene expression. Genome Res 20:554–564

Harrison CJ et al (2010) Detection of prognostically relevant genetic abnormalities in childhood B-cell precursor acute lymphoblastic leukaemia: recommendations from the Biology and Diagnosis Committee of the International Berlin-Frankfürt-Münster study group. Br J Haematol 151:132–142

Hennekam RC (2006) Hutchinson–Gilford progeria syndrome: review of the phenotype. Am J Med Genet A 140:2603–2624

Hewitt SL, High FA, Reiner SL, Fisher AG, Merkenschlager M (2004) Nuclear repositioning marks the selective exclusion of lineage-inappropriate transcription factor loci during T helper cell differentiation. Eur J Immunol 34:3604–3613

Holaska JM, Lee KK, Kowalski AK, Wilson KL (2003) Transcriptional repressor germ cell-less (GCL) and barrier to autointegration factor (BAF) compete for binding to emerin in vitro. J Biol Chem 278:6969–6975

Ishii K, Arib G, Lin C, Van Houwe G, Laemmli UK (2002) Chromatin boundaries in budding yeast: the nuclear pore connection. Cell 109:551–562

Kill IR (1996) Localisation of the Ki-67 antigen within the nucleolus. Evidence for a fibrillarin-deficient region of the dense fibrillar component. J Cell Sci 109:1253–1263

Klein IA, Resch W, Jankovic M, Oliveira T, Yamane A, Nakahashi H, Di Virgilio M, Bothmer A, Nussenzweig A, Robbiani DF, Casellas R, Nussenzweig MC (2011) Translocation-capture sequencing reveals the extent and nature of chromosomal rearrangements in B lymphocytes. Cell 147:95–106

Knight M, Ittiprasert W, Odoemelam EC, Adema CM, Miller A, Raghavan N, Bridger JM (2011) Non-random organization of the *Biomphalaria glabrata* genome in interphase Bge cells and the spatial repositioning of activated genes in cells co-cultured with *Schistosoma mansoni*. Int J Parasitol 41:61–70

Korfali N, Wilkie G, Swanson S, Srsen V, de las Heras J, Batrakou D, Malik P, Zuleger N, Kerr A, Florens L, Schirmer E (2012) The nuclear envelope proteome differs notably between tissues. Nucleus 3:552–564

Korostowski L, Raval A, Breuer G, Engel N (2011) Enhancer-driven chromatin interactions during development promote escape from silencing by a long non-coding RNA. Epigenetics Chromatin 4:21

Kosak ST, Skok JA, Medina KL, Riblet R, Le Beau MM, Fisher AG, Singh H (2002) Subnuclear compartmentalization of immunoglobulin loc. during lymphocyte development. Science 296:158–162

Koutna I, Kozubek S, Zaloudik J, Kozubek M, Lukasova E, Matula P, Bartova E, Skalnikova M, Cafourkova A, Jirsova P (2000) Topography of genetic loci in tissue samples: towards new diagnostic tool using interphase FISH and high-resolution image analysis techniques. Anal Cell Pathol 20:173–185

Kozubek S, Lukasova E, Ryznar L, Kozubek M, Liskova A, Govorun RD, Krasavin EA, Hornbeck G (1997) Distribution of ABL and BCR genes in cell nuclei of normal and irradiated lymphocytes. Blood 89:4537–4545

Kozubek S, Lukasova E, Mareckova A, Skalnikova M, Kozubek M, Bartova E, Kroha V, Krahulcova E, Slotova J (1999) The topological organization of chromosomes 9 and 22 in cell nuclei has a determinative role in the induction of t(9,22) translocations and in the pathogenesis of t(9,22) leukemias. Chromosoma (Berl) 108:426–435

Krystosek A (1998) Repositioning of human interphase chromosomes by nucleolar dynamics in the reverse transformation of HT1080 fibrosarcoma cells. Exp Cell Res 241:202–209

Kubben N, Adriaens M, Meuleman W, Voncken J, van Steensel B, Misteli T (2012) Mapping of lamin A- and progerin-interacting genome regions. Chromosoma (Berl) 121:447–464

Kuroda M, Tanabe H, Yoshida K, Oikawa K, Saito A, Kiyuna T, Mizusawa H, Mukai K (2004) Alteration of chromosome positioning during adipocyte differentiation. J Cell Sci 117:5897–5903

Lakshmikuttyamma A, Scott SA, DeCoteau JF, Geyer CR (2010) Reexpression of epigenetically silenced AML tumor suppressor genes by SUV39H1 inhibition. Oncogene 29:576–588

Li C, Shi Z, Zhang L, Huang Y, Liu A, Jin Y, Yu Y, Bai J, Chen D, Gendron C, Liu X, Fu S (2010) Dynamic changes of territories 17 and 18 during EBV-infection of human lymphocytes. Mol Biol Rep 37:2347–2354

Lukasova E, Kozubek S, Kozubek M, Kjeronska J, Ryznar L, Horakova J, Krahulcova E, Horneck G (1997) Localisation and distance between ABL and BCR genes in interphase nuclei of bone marrow cells of control donors and patients with chronic myeloid leukaemia. Hum Genet 100:525–535

Manuelidis L, Borden J (1988) Reproducible compartmentalization of individual chromosome domains in human CNS cells revealed by in situ hybridization and three-dimensional reconstruction. Chromosoma (Berl) 96:397–410

Marella NV, Bhattacharya S, Mukherjee L, Xu J, Berezney R (2009) Cell type specific chromosome territory organization in the interphase nucleus of normal and cancer cells. J Cell Physiol 221:130–138

Meaburn KJ, Misteli T (2008) Locus-specific and activity-independent gene repositioning during early tumorigenesis. J Cell Biol 180:39–50

Meaburn KJ, Cabuy E, Bonne G, Levy N, Morris GE, Novelli G, Kill IR, Bridger JM (2007a) Primary laminopathy fibroblasts display altered genome organization and apoptosis. Aging Cell 6:139–153

Meaburn KJ, Misteli T, Soutoglou E (2007b) Spatial genome organization in the formation of chromosomal translocations. Semin Cancer Biol 17:80–90

Meaburn KJ, Gudla PR, Khan S, Lockett SJ, Misteli T (2009) Disease-specific gene repositioning in breast cancer. J Cell Biol 187:801–812

Mehta IS, Figgitt M, Clements CS, Kill IR, Bridger JM (2007) Alterations to nuclear architecture and genome behavior in senescent cells. Ann N Y Acad Sci 1100:250–263

Mehta IS, Amira M, Harvey AJ, Bridger JM (2010) Rapid chromosome territory relocation by nuclear motor activity in response to serum removal in primary human fibroblasts. Genome Biol 11:R5

Mehta IS, Eskiw CH, Arican HD, Kill IR, Bridger JM (2011) Farnesyltransferase inhibitor treatment restores chromosome territory positions and active chromosomes dynamics in Hutchinson–Gilford progeria syndrome cells. Genome Biol 12:R74

Mewborn SK, Puckelwartz MJ, Abuisneineh F, Fahrenbach JP, Zhang Y, MacLeod H, Dellefave L, Pytel P, Selig S, Labno CM, Reddy K, Singh H, McNally E (2010) Altered chromosomal positioning, compaction, and gene expression with a lamin A/C gene mutation. PLoS One 5:e14342

Morgan R, Sandberg AA, Jarzabek V, Scheerer P, Hecht F (1987) An acquired Robertsonian translocation in prolymphocytic leukemia: a case presentation and review. Cancer Genet Cytogenet 25:293–301

Murata S, Nakazawa T, Ohno N, Terada N, Iwashina M, Mochizuki K, Kondo T, Nakamura N, Yamane T, Iwasa S, Ohno S, Katoh R (2007) Conservation and alteration of chromosome territory arrangements in thyroid carcinoma cell nuclei. Thyroid 17:489–496

Murmann AE, Gao J, Encinosa M, Gautier M, Peter ME, Eils R, Lichter P, Rowley JD (2005) Local gene density predicts the spatial position of genetic loci in the interphase nucleus. Exp Cell Res 311:14–26

Murphy KM, Reiner SL (2002) The lineage decisions of helper T cells. Nat Rev Immunol 2:933–944

Nagano T, Mitchell JA, Sanz LA, Pauler FM, Ferguson-Smith AC, Feil R, Fraser P (2008) The Air noncoding RNA epigenetically silences transcription by targeting G9a to chromatin. Science 322:1717–1720

Nemeth A, Conesa A, Santoyo-Lopez J, Medina I, Montaner D, Peterfia B, Solovei I, Cremer T, Dopazo J, Langst G (2010) Initial genomics of the human nucleolus. PLoS Genet 6:e1000889

Nikiforova MN, Stringer JR, Blough R, Medvedovic M, Fagin JA, Nikiforov YE (2000) Proximity of chromosomal loci that participate in radiation-induced rearrangements in human cells. Science 290:138–141

Noordermeer D, Leleu M, Splinter E, Rougemont J, De Laat W, Duboule D (2011) The dynamic architecture of Hox gene clusters. Science 334:222–225

Ondrej V, Lukasova E, Falk M, Kozubek S (2007) The role of actin and microtubule networks in plasmid DNA intracellular trafficking. Acta Biochim Pol 54:657–663

Ondrej V, Lukasova E, Krejci J, Kozubek S (2008a) Intranuclear trafficking of plasmid DNA is mediated by nuclear polymeric proteins lamins and actin. Acta Biochim Pol 55:307–315

Ondrej V, Lukasova E, Krejci J, Matula P, Kozubek S (2008b) Lamin A/C and polymeric actin in genome organization. Mol Cells 26:356–361

Osborne CS, Chakalova L, Mitchell JA, Horton A, Wood AL, Bolland DJ, Corcoran AE, Fraser P (2007) Myc dynamically and preferentially relocates to a transcription factory occupied by Igh. PLoS Biol 5:e192

Parada LA, McQueen PG, Munson PJ, Misteli T (2002) Conservation of relative chromosome positioning in normal and cancer cells. Curr Biol 12:1692–1697

Parada LA, McQueen PG, Misteli T (2004) Tissue-specific spatial organization of genomes. Genome Biol 5:R44

Peric-Hupkes D, Meuleman W, Pagie L, Bruggeman SW, Solovei I, Brugman W, Graf S, Flicek P, Kerkhoven RM, van Lohuizen M, Reinders M, Wessels L, van Steensel B (2010) Molecular maps of the reorganization of genome-nuclear lamina interactions during differentiation. Mol Cell 38:603–613

Rodríguez S, Eriksson M (2010) Evidence for the involvement of lamins in aging. Curr Aging Sci 3:81–89

Roix JJ, McQueen PG, Munson PJ, Parada LA, Misteli T (2003) Spatial proximity of translocation-prone gene loci in human lymphomas. Nat Genet 34:287–291

Ronkainen H, Kauppila S, Hirvikoski P, Vaarala MH (2010) Evaluation of myosin VI, E-cadherin and beta-catenin immunostaining in renal cell carcinoma. J Exp Clin Cancer Res 29:2

Schirmer EC, Florens L, Guan T, Yates JR III, Gerace L (2003) Nuclear membrane proteins with potential disease links found by substractive proteomics. Science 301:1380–1382

Sengupta K, Upender MB, Barenboim-Stapleton L, Nguyen QT, Wincovitch SM, Garfield SH, Difilippantonio MJ, Ried T (2007) Artificially introduced aneuploid chromosomes assume a conserved position in colon cancer cells. PLoS ONE 2(2):e199

Shimi T, Pfleghaar K, Kojima S, Pack CG, Solovei I, Goldman AE, Adam SA, Shumaker DK, Kinjo M, Cremer T, Goldman RD (2008) The A- and B-type nuclear lamin networks: microdomains involved in chromatin organization and transcription. Genes Dev 22:3409–3421

Somech R, Shaklai S, Geller O, Amariglio N, Simon AJ, Rechavi G, Gal-Yam EN (2005) The nuclear-envelope protein and transcriptional repressor LAP2beta interacts with HDAC3 at the nuclear periphery, and induces histone H4 deacetylation. J Cell Sci 118:4017–4025

Splinter E, de Wit E, Nora EP, Klous P, van de Werken HJG, Zhu Y, Kaaij LJT et al (2011) The inactive X chromosome adopts a unique three-dimensional conformation that is dependent on Xist RNA. Genes Dev 25:1371–1383

Starr DA (2009) A nuclear-envelope bridge positions nuclei and moves chromosomes. J Cell Sci 122:577–586

Stromme P, Mangelsdorf ME, Shaw MA, Lower KM, Lewis SM, Bruyere H, Lutcherath V, Gedeon AK, Wallace RH, Scheffer IE, Turner G, Partington M, Frints SG, Fryns JP, Sutherland GR, Mulley JC, Gecz J (2002) Mutations in the human ortholog of Aristaless cause X-linked mental retardation and epilepsy. Nat Genet 30:441–445

Sullivan GJ, Bridger JM, Cuthbert AP, Newbold RF, Bickmore WA, McStay B (2001) Human acrocentric chromosomes with transcriptionally silent nucleolar organizer regions associate with nucleoli. EMBO J 20:2867–2874

Szczerbal I, Bridger JM (2010) Association of adipogenic genes with SC-35 domains during porcine adipogenesis. Chromosome Res 18:887–895

Szczerbal I, Foster HA, Bridger JM (2009) The spatial repositioning of adipogenesis genes is correlated with their expression status in a porcine mesenchymal stem cell adipogenesis model system. Chromosoma (Berl) 118:647–663

Taddei A, Van Houwe G, Hediger F, Kalck V, Cubizolles F, Schober H, Gasser SM (2006) Nuclear pore association confers optimal expression levels for an inducible yeast gene. Nature (Lond) 441:774–778

Taimen P, Pfleghaar K, Shimi T, Moller D, Ben-Harush K, Erdos MR, Adam SA, Herrmann H, Medalia O, Collins FS, Goldman AE, Goldman RD (2009) A progeria mutation reveals functions for lamin A in nuclear assembly, architecture, and chromosome organization. Proc Natl Acad Sci U S A 106:20788–20793

Taslerova R, Kozubek S, Lukasova E, Jirsova P, Bartova E, Kozubek M (2003) Arrangement of chromosome 11 and 22 territories, EWSR1 and FLI1 genes, and other genetic elements of these chromosomes in human lymphocytes and Ewing sarcoma cells. Hum Genet 112:143–155

Timme S, Schmitt E, Stein S, Schwarz-Finsterle J, Wagner J, Walch A, Werner M, Hausmann M, Wiech T (2011) Nuclear position and shape deformation of chromosome 8 territories in pancreatic ductal adenocarcinoma. Anal Cell Pathol (Amst) 34:21–33

Tolhuis B, Palstra RJ, Splinter E, Grosveld F, de Laat W (2002) Looping and interaction between hypersensitive sites in the active beta-globin loc.s. Mol Cell 10:1453–1465

Trinkle-Mulcahy L, Lamond AI (2007) Toward a high-resolution view of nuclear dynamics. Science 318:1402–1407

van Koningsbruggen S, Gierlinski M, Schofield P, Martin D, Barton GJ, Ariyurek Y, den Dunnen JT, Lamond AI (2010) High-resolution whole-genome sequencing reveals that specific chromatin domains from most human chromosomes associate with nucleoli. Mol Biol Cell 21:3735–3748

Wiech T, Timme S, Riede F, Stein S, Schuricke M, Cremer C, Werner M, Hausmann M, Walch A (2005) Human archival tissues provide a valuable source for the analysis of spatial genome organization. Histochem Cell Biol 123:229–238

Wiech T, Stein S, Lachenmaier V, Schmitt E, Schwarz-Finsterle J, Wiech E, Hildenbrand G, Werner M, Hausmann M (2009) Spatial allelic imbalance of BCL2 genes and chromosome 18 territories in nonneoplastic and neoplastic cervical squamous epithelium. Eur Biophys J 38:793–806

Yanaihara N, Nishioka M, Kohno T, Otsuka A, Okamoto A, Ochiai K, Tanaka T, Yokota J (2004) Reduced expression of MYO18B, a candidate tumor-suppressor gene on chromosome arm 22q, in ovarian cancer. Int J Cancer 112:150–154

Zhang L, Lan Q, Guo W, Hubbard AE, Li G, Rappaport SM, McHale CM, Shen M, Ji Z, Vermeulen R, Yin S, Rothman N, Smith MT (2011) Chromosome-wide aneuploidy study (CWAS) in workers exposed to an established leukemogen, benzene. Carcinogenesis (Oxf) 32:605–612

Zhang Y, McCord RP, Ho YJ, Lajoie BR, Hildebrand DG, Simon AC, Becker MS, Alt FW, Dekker J (2012) Spatial organization of the mouse genome and its role in recurrent chromosomal translocations. Cell 148:908–921

Zhu Q, Pao GM, Huynh AM, Suh H, Tonnu N, Nederlof PM, Gage FH, Verma IM (2011) BRCA1 tumour suppression occurs via heterochromatin-mediated silencing. Nature (Lond) 477:179–184

Zink D, Fischer AH, Nickerson JA (2004) Nuclear structure in cancer cells. Nat Rev Cancer 4:677–687

Zuleger N, Robson MI, Schirmer EC (2011) The nuclear envelope as a chromatin organizer. Nucleus 2:339–349

Chapter 3
Nuclear Architecture, Chromosome Aberrations, and Genetic Damage

Gustavo A. Folle, María Vittoria Di Tomaso, Laura Lafon-Hughes, and Pablo Liddle

Abstract Mammalian interphase nuclei are highly organized structures in which chromosome territories are nonrandomly distributed following a radial pattern. Gene richness and size markedly influence nuclear chromosome positioning. Active euchromatin and inactive heterochromatin exhibit different nuclear topology: the former is centrally located and the latter mostly placed at the periphery, within chromocenters, and around nucleoli. DNA is subjected to a wide repertoire of insults including radiation, chemical, and biological agents. The differential sensitivity of euchromatin and heterochromatin to clastogens has been a matter of debate, although most experimental evidence supports that euchromatin is more damage prone. Gene expression and DNA synthesis coupled to chromatin remodeling could act as key factors in the distribution of chromosome aberrations (CA) in euchromatic and heterochromatic regions of genome. In this chapter, the main features of nuclear architecture as well as an overview of current knowledge of genetic damage at the metaphase and interphase levels are presented. Also, the preferential involvement of transcriptionally active regions of the human genome regarding the induction of chromosome aberrations and deregulation of tumor genes is analyzed. Finally, the impact of DNA replication timing and connected chromatin remodeling processes in the generation and localization of CA and primary genetic damage is discussed.

G.A. Folle (✉) • M.V. Di Tomaso • L. Lafon-Hughes • P. Liddle
Department of Genetics, Instituto de Investigaciones Biológicas, Clemente Estable (IIBCE), Av. Italia 3318 CP, 11600 Montevideo, Uruguay
e-mail: folle@iibce.edu.uy; marvi@iibce.edu.uy; laulaf@iibce.edu.uy; liddle@iibce.edu.uy

Y.B. Yurov et al. (eds.), *Human Interphase Chromosomes: Biomedical Aspects*,
DOI 10.1007/978-1-4614-6558-4_3,

Introduction

Human populations exhibit a significant frequency of inborn chromosomal aberrations (CA). Therefore, it is of utmost importance to understand the mechanisms underlying their origin and transmission to progenies (Miller and Therman 2001). CA are among the major biological endpoints of human exposure to ionizing radiation and genotoxic compounds. Hence, CA scoring is a key tool for biological dosimetry of radiation exposure or disclosing putative mutagenic/carcinogenic agents. Moreover, specific CA are associated with different types of cancers (Obe et al. 2002).

Early cytogenetic analyses based on classical solid staining of metaphase chromosomes provided valuable information on the types and frequencies of spontaneous and clastogen-induced CA. Exposure of cells during the G_0/G_1-phase leads to chromosome-type aberrations whereas chromatid-type lesions are induced after DNA replication (S/G_2 phases). Chromosome banding procedures allowed the precise recognition of individual chromosomes, karyotype evolutionary studies, and assignment of genes to specific chromosome landmarks as well as accurate mapping of chromosome breakpoints either induced by DNA-damaging agents or present in inborn or neoplastic diseases.

Moreover, banding techniques paved the way to understanding the highly complex structure of mammalian chromosomes. G- and R-banding revealed the presence of alternate Giemsa-dark and Giemsa-light chromosome bands, which reflect the evolutionary partition of euchromatin as well as constitutive and facultative heterochromatin into distinct chromosome domains. G-light bands are gene-rich, high G-C content, early-replicating regions, harboring short interspersed repeated elements (SINEs). Most housekeeping genes map to G-light bands and exhibit distinct epigenetic modifications such as unmethylated CpG islands and histone H3/H4 hyperacetylation. On the other hand, G-dark bands are gene-poor, late-replicating, A-T-rich regions, homing tissue-specific genes and long interspersed repeated elements (LINEs). C-band constitutive heterochromatin is highly packed, inactive, and almost devoid of genes. Epigenetic modifications in G-dark and C-bands include DNA and specific histone lysine methylation as well as serine H3/H4 underacetylation (Holmquist 1992; Korenberg and Rykowski 1988).

The impact of chromatin organization in the induction and localization of genetic damage induced by clastogens has been extensively studied by several research groups (Obe et al. 2002 and citations therein). The interchromosomal distribution of CA seems to be random (Cornforth et al. 2002; Martínez-López et al. 2000), although nonrandomness has also been claimed (Grigorova et al. 1998; Xiao and Natarajan 1999) analyzing either G-banded or FISH-painted chromosomes. In contrast, the intrachromosomal distribution of CA has been repeatedly reported to be nonrandom (Obe et al. 2002; Slijepcevic and Natarajan 1994). A higher susceptibility of G-light chromatin in Chinese hamster ovary (CHO) cells after electroporation of restriction endonucleases (RE) or DNAse I or even following irradiation with neutrons and γ-rays has been reported. Moreover, clustering of both endonuclease- and radiation-induced breakpoints in specific CHO chromosome regions independently of the cell-cycle stage (G_1- or S-phase) was also evidenced (Folle and Obe 1995, 1996; Folle et al. 1997, 1998).

Immunolabeling of hyperacetylated histone H4 ($H4^{+a}$) in CHO metaphase spreads revealed the colocalization of $H4^{+a}$ chromosome regions with endonuclease- and radiation-induced breakpoint clusters, giving further support to the assumption that transcriptionally active euchromatin is a preferential target for DNA damage induction (Martínez-López et al. 2001).

The detection of specific DNA sequences and even whole chromosomes (or chromosome arms) in metaphase spreads through fluorescence in situ hybridization (FISH) techniques has greatly improved our knowledge regarding the mechanisms involved in the formation of CA. FISH analyses uncovered CA types and mechanisms not evidenced by conventional staining or banding procedures (Simpson and Savage 1995). Molecular cytogenetic techniques such as spectral karyotyping (SKY), comparative genome hybridization (CGH), multicolor FISH (mFISH), and multicolor chromosome banding (mBAND FISH) precisely identify a wide spectrum of CA ranging from drastic genomic alterations to minute chromosome changes. Interphase cytogenetic studies based on FISH or immunoFISH methods have furnished crucial information concerning the organization and function of eukaryotic chromosomes and nuclei. Correlative metaphase and interphase cytogenetic studies are now possible, especially in relationship to the origin and fate of CA. Still, our present understanding of DNA damage processing leading to CA formation remains incomplete.

Nuclear Architecture

The eukaryotic nucleus is a highly organized and compartmentalized structure reflecting genome and epigenome dynamics at the cellular and molecular level. The first compaction level of chromatin is the 10-nm fiber involving DNA wrapped around nucleosomes. Further chromatin folding could lead to the 30-nm fiber and to still poorly known higher-order structures critical for the building and dynamics of eukaryotic nuclei. Chromatin compaction levels seem to play a critical role in nuclear organization. Highly packed constitutive and facultative heterochromatin (C- and G-dark bands) reside at the nuclear periphery, around nucleoli, and within chromocenters whereas euchromatin (G-light bands) is centrally located (Postberg et al. 2010). The position of interphase chromosomes inside mammalian nuclei is nonrandom and radially organized. Chromosome positioning is mainly modulated by size and gene density but not by gene expression (Küpper et al. 2007). However, chromosome nuclear location may vary depending on cell type and shape, differentiation state, presence of chromosomal aberrations, senescence, and transformation. Interphase chromosomes are structured as discrete, nonoverlapping territories (CTs), although intermingling of peripheral chromatin loops was also proposed (Branco and Pombo 2006). Gene-rich chromosomes are inwardly positioned and gene-poor chromosomes mostly lodge in the nuclear boundary (Cremer et al. 2003). This chromosome gene richness-dependent topological partition has been shown to be evolutionarily conserved (Tanabe et al. 2002).

It is noteworthy that plasticity seems to be also an inherent feature of nuclear architecture, as evidenced by striking nuclear multilobation in granulocytes, and manifold modifications in cancer cells, as well as the striking relocation of euchromatin and heterochromatin in retina rod cells to improve light collection in nocturnal species (Solovei et al. 2009).

A central role of nuclear chromatin assembly could be the configuration of two distinct functional domains: condensed silent heterochromatin and transcriptionally active euchromatin. Packed heterochromatin shows low acetylating levels as well as methylated DNA and H3K9/H3K27 histone tail methylation whereas open euchromatin is highly acetylated and enriched in H3K4me and H3K79me. Nuclear chromatin compartments also influence gene expression and DNA synthesis patterns. In fact, transcription dominates in central regions although it is nearly absent from the nuclear rim (Sadoni et al. 1999, 2004). Furthermore, transcriptional hubs could be very significant for high throughput transcription and the expression of coregulated genes. DNA synthesis progression from early-replicating gene-rich to mid-/late-replicating gene-poor chromatin compartments is also highly ordered (see following).

According to the CT/IC model, CTs are bordered by the interchromatin compartment (IC), which harbors nuclear bodies and the complex machineries required for DNA transcription, replication, and repair as well as for RNA processing. The IC comprise an intricate channel network that expands throughout the nucleus, reaching the interior of CTs and the nuclear pores. The interface between CTs and the IC, named the perichromatin region (PR), has been proposed as a key nuclear subcompartment for DNA/RNA metabolism (Postberg et al. 2010; Rouquette et al. 2009).

DNA Damage and Chromosome Aberrations: The Nuclear Landscape

The DNA double-strand break (DSB) is the ultimate lesion leading to chromosomal aberrations, genomic instability, oncogenic transformation, and cell death (Obe et al. 2002). DSB may arise as a result of endogenous oxidative damage by oxygen free radicals or collapsed replication forks during DNA synthesis. A wide spectrum of physical, chemical, and biological DNA-damaging agents are also able to induce DSB. Two main repair systems evolved to cope with these deleterious and highly recombinogenic DNA lesions: nonhomologous end-joining (NHEJ), mostly active during the G_1 stage of the cell cycle, and homologous recombination (HR), which acts along late S/G_2-phases and requires an undamaged template for efficient DSB repair (Rogakou et al. 1998). DSB mobility is constrained within the nuclear volume and normally is not beyond 0.5 μm, which corresponds to the Brownian motion radius in cells (Jakob et al. 2009).

DNA damage triggers a global response in cells named DNA damage response (DDR), which activates a complex mechanism committed to detect lesions and restore DNA integrity. The induction of DSB elicits the phosphorylation of the variant histone H2AX on serine 139 (γH2AX) by PIKK (phosphoinositide 3-kinase-related kinase) ATM. Other PIKKs such as ATR and DNA-PKcs may also phosphorylate histone H2AX. Soon after DSB induction (~3 min), γH2AX molecules are generated over megabase chromatin domains flanking the lesion. A plateau in the yield of γH2AX foci is reached approximately 30 min after damage induction. In addition to signaling the presence of DSB, γH2AX molecules interact and may provide docking sites for repair proteins including the MRE11/NSB1/RAD50 complex, MCD1, 53BP1, and BRCA1. The H2AX phosphorylation response can be unveiled as discrete foci through immunodetection with specific antibodies allowing the quantitation of primary DNA lesions in interphase nuclei. As H2AX phosphorylation is proportional to DSB induction, γH2AX is considered a sensitive biomarker of DNA damage (Podhorecka et al. 2010). However, the presence of spontaneous γH2AX foci in human confluent fibroblast cultures and senescent cells as well as in aged mice tissues has been reported. S-phase cells exhibit distinct punctuate small γH2AX foci (Costes et al. 2010 and citations therein).

The distribution of γH2AX foci in the different nuclear compartments after DNA insult has been extensively studied. A preferential localization of γH2AX foci in the nuclear interior after treating human fibrosarcoma H1080 cells with UV radiation or hydrogen peroxide was observed. XRCC repair proteins also mapped to the same nuclear domains (Gazave et al. 2005). Modulation of γH2AX foci distribution by chromatin density was also reported by Costes et al. (2010). In this study, foci located mainly in DAPI-weak regions (euchromatin) or at eu- and heterochromatin interfaces. Moreover, exclusion of γH2AX foci from heterochromatin was also observed in MCF7 breast carcinoma cells treated with either X-rays or the topoisomerase II inhibitor etoposide. Combined immunodetection of HP1 (heterochromatin protein 1) evidenced no colocalization with γH2AX foci. Comparable results were obtained when the heterochromatin marker H3K9me3 was assayed. Interestingly, γH2AX foci mapped to heterochromatin domains and colocalized with HP1 when MCF7 cells were treated with hydroxyurea during late S-phase (heterochromatin replication). In this case, DSB stem from stalled or collapsed replication forks and ATR signaling. It is envisaged that chromatin remodeling associated to DNA synthesis could render heterochromatin amenable to histone H2AX phosphorylation (Cowell et al. 2007). Finally, it has been shown that specific heterochromatic regions such as the alpha satellite and satellite 2 are resistant to the formation of γH2AX foci after irradiation. Still, pretreatment of cells with the deacetylase inhibitor trychostatin A (TSA) causes these satellite sequences to be prone to generate γH2AX foci (Karagiannis et al. 2007).

These findings argue in favor of a higher sensitivity of euchromatin to DNA insult. On the other hand, they challenge the hypothesis of a protective role of the heterochromatin compartment adjacent to the nuclear envelope by shielding the gene-rich central region. There is yet no clear-cut explanation to heterochromatin low sensitivity to γH2AX foci formation, although the following factors have been

Table 3.1 Molecular size of selected gene-rich (HSA11, HSA19), gene-poor (HSA4, HSA18), and low gene density (HSA2) human chromosomes (Mbp) with their corresponding number of genes and radiation-induced DNA double-strand breaks (DSB) per Mbp estimated via γH2AX foci

Chromosome number	Mbp	Genes per Mbp	DSB per Mbp[a]	Number of TDRG[b]	TDRG per Mbp
2	243.6	6.2	0.05	68	0.28
4	191.7	4.8	0.03	37	0.19
11	134.4	13.75	0.07	52	0.39
18	76.1	4.3	0.03	12	0.16
19	63.8	23.9	0.12	60	0.94

Respective amounts of tumor-deregulated genes (TDRG) per chromosome and per Mbp are also depicted
[a]Data from Falk et al. (2008)
[b]Data from Folle et al. (2010)

postulated: (a) low level of histone H2AX; (b) nominal H2AX phosphorylation; (c) scarce induction of DSB because of high chromatin compaction; (d) rapid migration of genuine heterochromatic DSB to neighboring euchromatic domains; and (e) loss of heterochromatin features as a result of DSB-induced local chromatin decondensation (Cowell et al. 2007). Experimental evidence highlights the impact of chromatin relocation after DNA damage. H2AX is effectively phosphorylated in chromocenters after single-ion microbeam irradiation, but the damaged region is rapidly decondensed and shortly expelled to the chromocenter periphery (Jakob et al. 2011).

Cytogenetic evidence on the role played by chromatin structure in the localization of chromosome breakpoints induced by DNA-damaging agents (Folle 2008) has been confirmed through immunoFISH studies in nuclei of human fibroblasts (Falk et al. 2008). Results obtained indicate that genetically inactive condensed chromatin is less susceptible to DSB induction by γ-rays than are transcriptionally active regions. The amount of γH2AX foci produced by ionizing radiation in the CTs of gene-dense (HSA11, HSA19), intermediate density (HSA2), and gene-poor (HSA4, HSA18) human chromosomes was assessed. Gene-dense CTs exhibited higher yields of DNA DSB (visualized as γH2AX foci) per Mbp than intermediate or gene-poor CTs (Table 3.1).

Interestingly, damage-prone HSA11 and HSA19 harbor more tumor-deregulated genes (TDRG) per Mbp than gene-poor HSA4 and HSA18 whereas HSA 2 exhibits an intermediate yield (Folle et al. 2010) (Table 3.1). In other words, the chromosomal extent of TDRG is in agreement with damage sensitivity to ionizing radiation (see following).

The nonrandom organization of CTs within eukaryotic nuclei could influence the pattern of chromosome aberration production. Hence, nonrandom occurrence of chromosome rearrangements is expected to occur because vicinity effects and constrained migration of chromatin loops may limit the range of putative partner domains.

A good example in this respect is given by chicken DT40 (lymphocytes) and CEF (embryonic fibroblasts) cell lines. DT40 cells typically exhibit chicken gene-rich minichromosomes (MICs) in the nuclear interior whereas gene-poor macrochromosomes (MACs) dwell at the periphery (Habermann et al. 2001). Irradiated DT40 cells exhibited mainly MIC/MIC and MAC/MAC translocations. The topological partition of MICs and MACs in chicken nuclei is lost in CEF cells (MICs distribute throughout the nuclear volume), favoring neighborhoods between both chromosome types. As expected, CEF cells showed an increased frequency of MIC/MAC translocations (Grandy et al. 2002).

As mentioned earlier, CT organization and neighborhoods may differ between cell lineages, and thus variations in chromosome translocation frequencies are also expected to happen. In mouse lymphomas CT 12 and 15 are close neighbors whereas in mouse hepatomas CT 5 and 6 are topologically related. Analysis of translocation rates revealed that 5/6 rearrangements frequently occur in hepatomas but are absent in lymphomas. Conversely, 12/15 translocations prevail in lymphomas. In conclusion, the differential spatial organization of genomes in specific tissues could be critical for the formation of recurrent CA (Parada et al. 2004).

According to vicinity effects and DSB-constrained mobility, a higher frequency of chromosome rearrangements within CTs is expected to occur. However, most data point to a ratio of inter/intrachromosomal CA (F ratio) above 1; that is, an excess of interchromosome exchanges is regularly detected even for high linear energy transfer (LET) radiation where DSB are formed in close proximity along the tracks. These findings have been confirmed using high-resolution mBAND FISH and densely ionizing radiation (Johannes et al. 2004; Obe and Durante 2010). The basis of this unpredicted distribution of CA remains to be elucidated.

Recurrent translocations in Burkitt's lymphoma involve the *MYC* locus at 8q24 with different partners encoding immunoglobulin heavy chain (*IGH*, 14q21), light chain α (*IGL*, 22q11) and light chain κ (*IGK*, 2p11) gene loci. Translocation frequencies differ, being *MYC-IGH*> *MYC-IGL*> *MYC-IGK*. FISH detection of *MYC* and immunoglobulin loci with specific probes in karyotypically normal MC/CAR cells revealed a direct relationship between translocation frequencies and nuclear spatial proximity to the *MYC* locus (Roix et al. 2003). *MYC* deregulation results in nuclear remodeling of telomere and chromosome positions, preceding the onset of chromosomal rearrangements and instability (Louis et al. 2005). Spatial vicinity has been claimed as a decisive factor for some cancer-prone loci rearrangements (Gué et al. 2006; Weckerle et al. 2011) as well as for the coregulation of genes (Brown et al. 2006).

The impact of chromosome translocations on gene expression and CT repositioning has been well documented by Harewood et al. (2010). Transcriptomes pertaining to balanced translocation t(11;22)(q23;q11) carriers, Emanuel syndrome patients with +der(22)t(11;22)(q23;q11) unbalanced karyotypes, as well as from normal individuals were analyzed. Significant variations in the number of differentially expressed transcripts (DET) between translocation-carrying and normal cohorts were observed. Because Emanuel syndrome patients are partially trisomic for chromosomes 11 and 22, the expected increase in DET mapping to these regions was confirmed. In the case of balanced t(11;22)(q23:q11), many DET corresponded to genes located along the derived chromosome 11. The modification of gene expression in translocation carriers involved not only the breakpoint region but also genes residing tens of megabases apart or mapping to the opposite arm or other chromosomes. FISH studies revealed a spatial reorganization of derived chromosome 11 toward the nuclear center compared to its normal counterpart. However, no significant changes in the position of derived 22 were observed. Anchoring of NOR regions to the nucleolus could explain the constrained mobility of derived 22. It is noteworthy that a displacement of gene-rich HSA17 to a more peripheral nuclear environment was observed in translocation carriers, reflecting long-range CT modifications as a consequence of derived 11 repositioning. The authors conclude that karyotype changes may lead to large-scale variations in gene expression and alterations in CT positioning, which could impinge on genome instability, tumor development, and speciation.

Tissue-specific CT organization may drastically change in malignant cells. Variations in the location of gene-poor HSA18 and gene-rich HSA19 was observed in malignant cell lines characterized by nuclei showing inverted positional patterns and a decline in gene density nuclear order. Interestingly, positional changes of HSA18 and HSA19 have also been reported in colon carcinoma cell lines RKO and DLD1, which exhibit nearly diploid karyotypes (Cremer et al. 2003).

Chromatin Dynamics: Gene Expression and Genetic Damage

Transcriptionally competent and actively transcribed chromatin maps to the nuclear interior and roughly corresponds to domains of early-replicating, gene-rich G-light bands. Similarly, hyperacetylated histone H4 ($H4^{+a}$), a cytogenetic marker for gene expression, is also confined to the inner compartment of mammalian nuclei. These facts reveal a higher-order spatial organization of functional nuclear processes (Jeppesen and Turner 1993; Sadoni et al. 1999).

The analysis of the human transcriptome map (HTM) disclosed the presence of regions of increased gene expression (RIDGEs) as well as genome domains of very low or null transcription (antiRIDGEs). RIDGEs typically are 5- to 15-Mbp gene-dense regions exhibiting high G-C content, short introns, and SINE repeats. AntiRIDGEs, in contrast, are gene-poor, low G-C content regions that harbor longer introns and LINE repeats. Moreover, RIDGEs show open chromatin conformation (Gilbert et al. 2004) and map to the nuclear interior whereas antiRIDGEs are condensed regions adjacent to the nuclear envelope (Caron et al. 2001; Gierman and Indemans 2007).

In an attempt to disclose the putative relationship between active chromatin ($H4^{+a}$, RIDGEs) susceptibility to DNA damage induction and tumor gene deregulation, we mapped, onto human G-band idiograms: (1) radiation breakpoint clusters (RBPC; $n=69$) (Barrios et al. 1989); (2) $H4^{+a}$ regions (Jeppesen 1997); (3) the transcriptome map (Caron et al. 2001); (4) RIDGEs and antiRIDGEs (Gierman and

Indemans 2007) and; (5) tumor-deregulated genes (TDRG; $n = 1019$) (Aouacheria et al. 2006) (Fig. 3.1a, b).

TDRG cluster at 23 chromosome regions characterized by the presence of RIDGEs, $H4^{+a}$ chromatin, and RBPC. Additionally, TDRG map to 12 regions harboring RIDGEs and enriched in $H4^{+a}$ but devoid of RBPC (see 14qcen, 16p, 17q, and Xqter). Note that some chromosome segments flanking RIDGEs also concentrate TDRG (e.g., 1p, 6p, 12p, and 19q; not shown). Conversely, fewer RBPC and TDRG mapped to antiRIDGEs. RBPC exhibiting no colocalization with RIDGEs or antiRIDGEs mostly corresponded to chromosome regions of medium to high gene expression (Folle et al. 2010).

However, because there is a good correlation between the HTM and gene density maps along human chromosomes, the observed clustering of TDRG in RIDGEs could respond, at least in part, to gene content variations between gene-rich and gene-poor genome regions. Through in silico methods it was possible to determine specific chromosome subbands implicated in TDRG independent of gene concentration (Aouacheria et al. 2006). As can be seen in Table 3.2, two thirds of these subbands map to RIDGEs, harbor RBPC, and participate in amplification/deletion events in tumor formation (antiRIDGEs = 0). Nearly all subbands (32 of 33) are embedded in $H4^{+a}$ chromatin and are sites of chromosome changes during evolution.

The higher sensitivity of RIDGEs compared to antiRIDGEs in relationship to genetic damage was also observed by Falk et al. (2008). In this study, the presence of radiation-induced γH2AX foci was determined in a RIDGE (11 Mbp) as well as in a nearby antiRIDGE (11 Mbp), both mapping to the pericentric region of 11q and only 12 Mbp apart. Similar to our findings, γH2AX foci were much more frequent (~4×) in the 11q RIDGE than in the antiRIDGE in which chromatin is nearly 40 % more condensed (Goetze et al. 2007).

The concentration of RBPC and TDRG in RIDGEs and $H4^{+a}$ chromatin could be the result of a compartment-biased response to DNA insult in highly expressed regions of the human genome. In this respect, spatial CT positioning (and repositioning), neighborhoods, and intermingling of specific DNA sequences as well as long-range interactions and coexpression of genes at transcription hubs could all play a role in the origin and location of recurrent chromosomal aberrations, gene deregulation, tumor development, and karyotype evolution.

Chromatin Dynamics: DNA Replication and Genetic Damage

In mammals, replication dynamics reflects the compartmentalized structure of interphase nuclei. Replication foci (RF) can be disclosed through the incorporation of halogenated nucleotides or 5-ethynyl-2′-deoxyuridine (EdU) to S-phase cells. The progression from early to mid- and finally late DNA synthesis along the different nuclear compartments has been subdivided into five main stages: (1) RF map to the nuclear interior; (2) the pattern of RF flows to the nuclear periphery although few

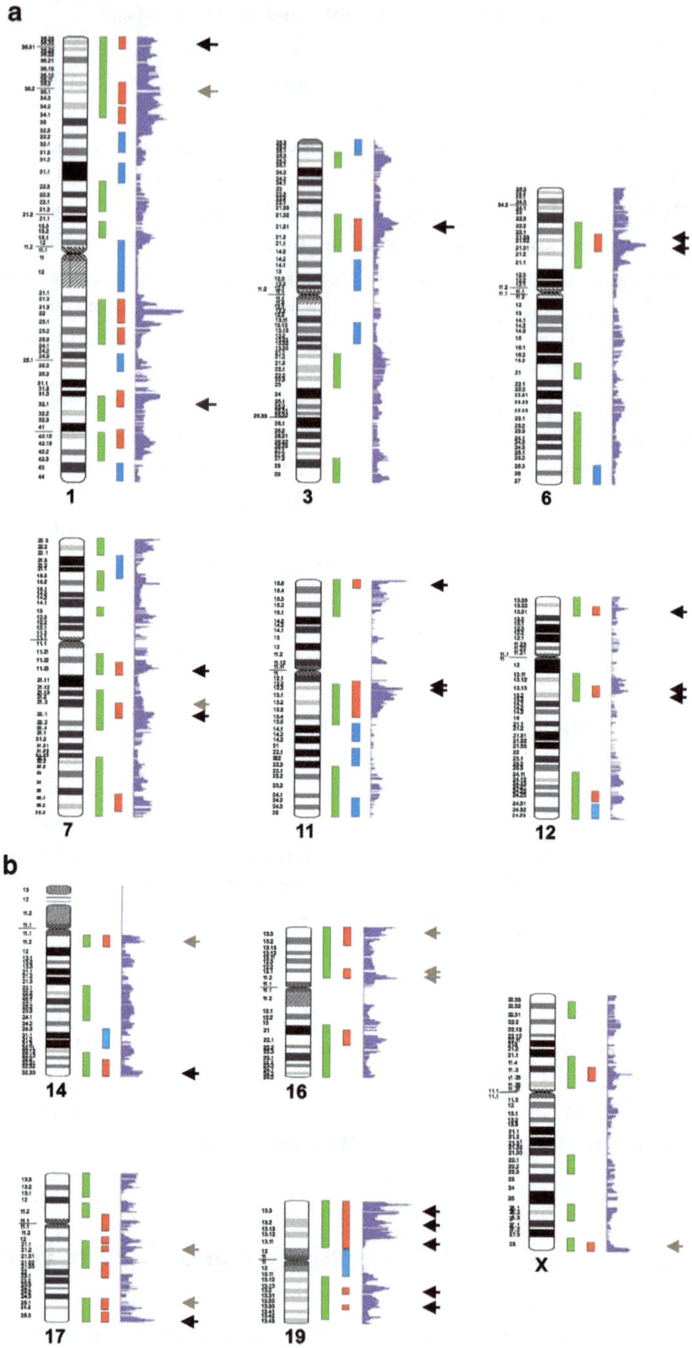

Fig. 3.1 Human idiograms depict chromosome regions harboring enriched hyperacetylated histone H4 ($H4^{+a}$, *green bars*), regions of increased gene expression (RIDGEs) (*red bars*), or anti-RIDGEs (*blue bars*), and the corresponding transcriptome map (*light violet*). (**a**) Chromosomes 1, 3, 6, 7, 11, and 12. (**b**) Chromosomes 14, 16, 17, 19, and X. *Black arrows* indicate colocalization of $H4^{+a}$ chromatin with RIDGEs, clusters of radiation breakpoints (Barrios et al. 1989), and clusters of tumor-deregulated genes (TDRG) (Aouacheria et al. 2006). *Gray arrows* denote colocalization of only $H4^{+a}$, RIDGEs, and TDRG. (Modified from Folle et al. 2010)

Table 3.2 Colocalization of clusters of tumor-deregulated genes (TDRGs) with regions of increased gene expression (RIDGEs), antiRIDGEs, H4[+a] chromatin, radiation breakpoint clusters (RBPC), amplification/deletion events in tumorigenesis (AMP/DEL), and breakpoints involved in evolution (EBP)

Chromosome (number of subbands)	RIDGEs	AntiRIDGEs	H4[+a]	RBPC	AMP/DEL	EBP
1 (2)	2	0	2	2	2	2
2 (3)	0	0	3	2	0	3
3 (1)	1	0	1	1	1	1
4 (1)	0	0	1	1	0	1
5 (1)	0	0	1	0	1	1
6 (1)	1	0	1	1	1	1
7 (3)	3	0	3	2	3	3
9 (2)	1	0	2	2	2	2
11(4)	2	0	4	4	3	4
12 (2)	1	0	2	2	1	2
14 (2)	2	0	2	1	2	2
16 (2)	1	0	1	0	2	2
17 (1)	1	0	1	0	1	1
19 (5)	5	0	5	4	4	5
20 (1)	0	0	1	1	0	1
21(1)	1	0	1	0	1	1
X (1)	1	0	1	0	1	0
17 (33)	*22*	*0*	*32*	*23*	*25*	*32*

A total of 33 chromosome subbands (in 17 chromosomes) that are implicated in tumor gene deregulation independently of gene density (Aouacheria et al. 2006) are listed (see text)

inner RF persist; (3) RF map only to the nuclear border and perinucleolar regions; (4) large and fewer RF are found at the nuclear rim and central region; (5) in late S-phase cells, large RF predominate in chromocenters residing in the nuclear interior and smaller RF map to peripheral heterochromatin domains (O'Keefe et al. 1992; Sadoni et al. 1999, 2004).

Replicating chromatin has been reported to be more sensitive to DNA-damaging agents (Cowell et al. 2007). Remodeling complexes associated to DNA synthesis could enhance chromatin accessibility and favor the induction of DNA damage in chromosome regions undergoing replication (Di Tomaso et al. 2010).

An interesting model to test the role played by DNA replication and chromatin remodeling in the production of chromosome breakpoints in mammalian cells has been developed by our group (Di Tomaso et al. 2006). The model takes advantage of the peculiar chromatin organization of the X chromosome in CHO cells. The short arm (Xp) is entirely euchromatic whereas heterochromatin is confined to the long arm (Xq) with the exception of a medial conspicuous secondary constriction (Xq_{sc}). Xp (as well as Xq_{sc}) exhibit hyperacetylated chromatin whereas Xq is underacetylated (Martínez-López et al. 2001). Pulse-labeling of CHO cultures with the base analogue bromodeoxyuridine (BrdU) allowed us to precisely define in metaphase spreads the S-phase stage (early, mid, or late) in which DNA insult occurred by BrdU immunodetection with fluorochrome-tagged specific antibodies. As one would expect, BrdU labeling is restricted to Xp during early S-phase and to Xq in late S-phase, whereas a combination of both patterns is observed in cells labeled during mid S-phase. By mapping chromosome breakpoints along CHO Xp and Xq in cells damaged and labeled during early (ES) or late (LS) S-phase, it is possible to assess the influence of replication timing and remodeling processes in BP localization.

Topoisomerase II (Topo II) modulates supercoiling and alleviates torsional stress in DNA through the generation of short-lived DSB that abrogate knots and tangles (Podhorecka et al. 2010). Topo II inhibitors (i.e., etoposide) are among the most effective anticancer drugs and potent inducers of DSB by stabilizing the cleavable complex of Topo II with DNA (Palitti 1993). Rolling DNA replication forks may collide with drug-stabilized Topo II–DNA complexes, turning transient DSB produced by Topo II into permanent DSB, leading to the induction of chromosomal aberrations.

Analysis of CA induced by etoposide in ES cells showed a concentration of BP in Xp. Conversely, clustering of BP in Xq was observed in cells exposed during LS (Fig. 3.2). A similar partition in the distribution of BP in Xp and Xq according to replication time was observed when the alkylating agent methyl methanesulfonate (MMS) was assayed (Di Tomaso et al. 2006). Exposure of CHO cells to UV radiation also showed BP clustering in Xp and Xq during ES and LS, respectively (Di Tomaso et al. 2010).

However, BP clustered in Xp but not in Xq after electroporation of the restriction endonuclease *Alu*I (5′-AGCT-3′) into ES and LS CHO cells, respectively (Di Tomaso et al. 2010). In this case, it can be argued that accessibility for bulky molecules such as *Alu*I (38 kDa) could still be hampered in replicating Xq heterochromatin during late S-phase. Supporting this view is the fact that Xq is highly resistant to *Alu*I digestion in metaphase spreads of CHO cells (Folle et al. 1997).

As mentioned, synchronized MCF7 cell cultures treated with hydroxyurea showed a preferential localization of γH2AX foci in heterochromatin during LS as evidenced by colocalization with HP1. Similar findings were reported when MCF7 cells were X-irradiated or exposed to the anticancer compound cisplatin (Cowell et al. 2007). Interestingly, an increment of γH2AX foci number was also observed in γ-irradiated human BJ skin fibroblast nuclei pretreated with hypotonic culture medium (~140 mOsm) to induce chromatin decondensation (Falk et al. 2008).

The drift of DNA replication along the different nuclear compartments allows studying the spatiotemporal relationship between RF (EdU labeling) and induced DSB (γH2AX foci immunodetection) in interphase nuclei.

The localization of RF and DSB (γH2AX foci) in CHO cells induced by the radiomimetic agent bleomycin in different stages of the S-phase is shown in Fig. 3.3. Some γH2AX foci colocalize with RF while others map to the boundaries of replication compartments. Most γH2AX foci do not map to EdU-unlabeled nuclear regions in either early/mid or late S-phase. These results underscore the impact of DNA synthesis and chromatin decondensation regarding the topology of DNA damage induction.

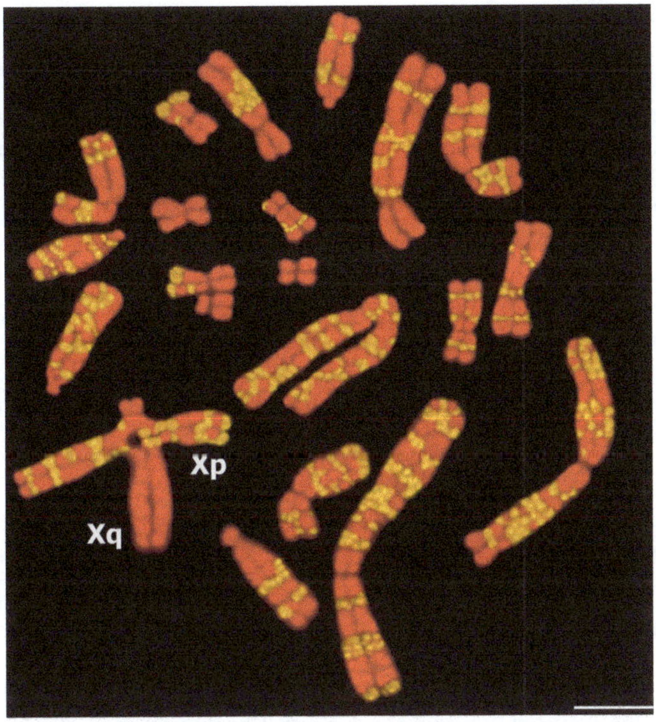

Fig. 3.2 CHO9 metaphase depicting early-replicating chromosome regions by immunodetection of BrdU incorporation using FITC-tagged anti-BrdU antibodies (*yellow*). Chromosomes were counterstained with propidium iodide (*red*). A quadrirradial rearrangement involving Xp and an acrocentric autosome is illustrated. Note the absence of replication labeling in Xq. *Bar* 5 μm

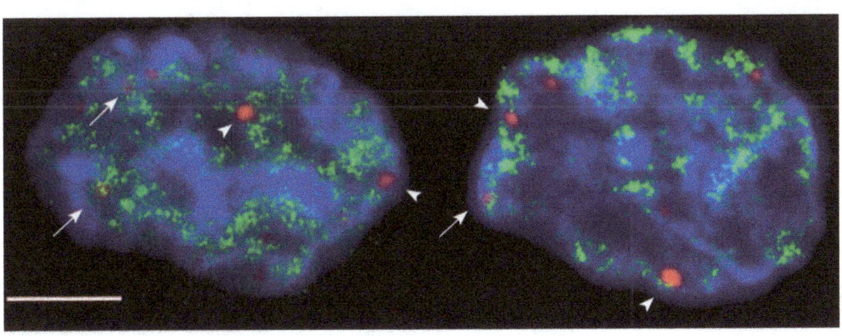

Fig. 3.3 Confocal z-sections showing the distribution of primary DNA damage (double-strand breaks, DSB) induced by bleomycin in CHO9 nuclei revealed as γH2AX foci with Cy3-bound specific antibodies (*red*) in relationship to replication foci (RF) detected with the EdU-Click-iT Alexa Fluor 488 Kit (*green*). An early/mid S-phase nucleus (*left*) and a late-replicating nucleus (*right*) are depicted. Nuclei were counterstained with DAPI. γH2AX foci positioning follows replication labeling. Colocalization of γH2AX foci with RF clusters is observed (*arrows*). γH2AX foci also map to the borders of replicating chromatin domains (*arrowheads*). (From Liddle P, unpublished results). *Bar* 3 μm

Recent findings involving the origin of somatic copy number alterations (SCNA) have shed new light concerning the participation of DNA replication in the mutational landscape of human cancer genomes. SCNA are abnormal structural variations of the genome arising through deletions or amplifications of chromosome segments. The study by De and Michor (2011) integrated databases from more than 330,000 SCNA boundaries inferred from microarray analysis of 2,792 samples pertaining to 26 different types of cancer, genome-wide DNA replication timing, and long-range (Hi-C) DNA interactions. They were able to demonstrate that SCNA arise preferentially in neighboring genomic regions (enriched for interactions in the Hi-C database) that share similar replication timing. Remarkably, early replicating regions exhibited more amplification SCNA events whereas deletions were more frequent in late-replicating domains. Thus, the spatiotemporal dynamics of DNA replication seems to play a crucial role in the generation of cancer-prone genomic rearrangements.

Conclusions

Nuclear structure and dynamics seem to impinge in the genesis and localization of CA. Genome regions with open chromatin conformation, high gene expression, or undergoing DNA replication constitute preferential sites for the induction of CA, thus paving the way for tumorigenesis. Spatial proximity may also be critical to determine partnerships in spontaneous or induced chromosome rearrangements. Correlative metaphase/interphase molecular cytogenetics has still much to offer to shed new light in the fields of chromatin organization and function, DNA damage handling, and CA production. New technological achievements that allow us to scrutinize the nuclear structure at the nanometer scale will certainly afford a new perspective to our present knowledge on the mechanisms of the origin of CA.

Acknowledgments We are indebted to PEDECIBA Postgraduate Program, the Agencia Nacional de Investigación e Innovación (ANII), and the Alexander von Humboldt-Foundation (AvH). P.L. is a former Fellow of an AvH Linkage Program research project.

References

Aouacheria A, Navratil V et al (2006) Bioinformatic screening of human ESTs for differentially expressed genes in normal and tumor tissues. BMC Genomics 7:94. doi:10.1186/1471-2164-7-94

Barrios L, Miró R et al (1989) Cytogenetic effects of radiotherapy breakpoint distribution in induced chromosome aberrations. Cancer Genet Cytogenet 41:61–70

Branco MR, Pombo A (2006) Intermingling of chromosome territories in interphase suggests a role in translocation and transcription-dependent associations. PLoS Biol 4:780–788

Brown JM, Leach J et al (2006) Coregulated human globin genes are frequently in spatial proximity when active. J Cell Biol 172(2):177–187

Caron H, van Schaik B et al (2001) The human transcriptome map: clustering of highly expressed genes in chromosomal domains. Science 291:1289–1292

Cornforth MN, Greulich-Bode M et al (2002) Chromosomes are predominantly located randomly with respect to each other in interphase human cells. J Cell Biol 159:237–244

Costes SV, Chiolo I et al (2010) Spatiotemporal characterization of ionizing radiation induced DNA damage foci and their relation to chromatin organization. Mutat Res 704:78–87

Cowell IG, Sunter NJ et al (2007) γH2AX form preferentially in euchromatin after ionising-radiation. PLoS One 2:e1057

Cremer M, Küpper K et al (2003) Inheritance of gene density-related higher order chromatin arrangements in normal and tumor cell nuclei. J Cell Biol 5:809–820

De S, Michor F (2011) DNA replication timing and long-range DNA interactions predict mutational landscapes of cancer genome. Nat Biotechnol 29(12):1103–1108. doi:10.1038/nbt.2030

Di Tomaso MV, Martínez-López W, Folle GA, Palitti F (2006) Modulation of chromosome damage localisation by replication timing. Int J Radiat Biol 82:877–886

Di Tomaso MV, Martínez-López W, Palitti F (2010) Asynchronously replicating eu/heterochromatic regions shape chromosome damage. Cytogenet Genome Res 128(1–3):111–117

Falk M, Lukášová E, Kozubek S (2008) Chromatin structure influences the sensitivity to γ-radiation. Biochim Biophys Acta 1783:2398–2414. doi:10.1016/j.bbamcr.2008.07.010

Folle GA (2008) Nuclear architecture, chromosome domains and genetic damage. Mutat Res 658:172–183. doi:10.1016/j.mrrev.2007.08.005

Folle GA, Obe G (1995) Localization of chromosome breakpoints induced by *Alu*I and *Bam*HI in Chinese hamster ovary (CHO) cells treated in G_1 phase of the cell cycle. Int J Radiat Biol 68:437–445

Folle GA, Obe G (1996) Intrachromosomal localization of breakpoints induced by the restriction endonucleases *Alu*I and *Bam*HI in Chinese hamster ovary cells treated in S-phase of the cell cycle. Int J Radiat Biol 69:447–457

Folle GA, Boccardo E, Obe G (1997) Localization of chromosome breakpoints induced by DNase I in Chinese hamster ovary (CHO) cells. Chromosoma (Berl) 106:391–399

Folle GA, Martínez-López W, Boccardo E, Obe G (1998) Localization of chromosome breakpoints: implication of the chromatin structure and nuclear architecture. Mutat Res 404:17–26

Folle GA, Liddle P, Lafon-Hughes L, Di Tomaso MV (2010) Close encounters: RIDGEs, hyperacetylated chromatin, radiation breakpoints and genes differentially expressed in tumours cluster at specific human chromosome regions. Cytogenet Genome Res 128(1–3):17–27

Gazave E, Gautier P, Gilchrist S, Bickmore WA (2005) Does radial nuclear organization influence DNA damage? Chromosome Res 13:377–388

Gierman HJ, Indemans MHJ (2007) Domain-wide regulation of gene expression in the human genome. Genome Res 17:1286–1295

Gilbert N, Boyle S et al (2004) Gene-rich domains are enriched in open chromatin fibers. Cell 118:555–566

Goetze S, Mateos-Langerak J et al (2007) The three-dimensional structure of human interphase chromosomes is related to the transcriptome map. Mol Cell Biol 27:4475–4487

Grandy I, Hardt T, Schmid M, Haaf T (2002) Effects of high-order nuclear structure and Rad51 overexpression on radiation-induced chromosome rearrangements. Cytogenet Genome Res 98:265–269. doi:10.1159/00007 1046

Grigorova M, Brand R, Xiao Y, Natarajan AT (1998) Frequencies and types of exchange aberrations induced by X-rays and neutrons in Chinese hamster splenocytes detected by FISH using chromosome-specific libraries. Int J Radiat Biol 74:297–314

Gué M, Sun J-S, Boudier T (2006) Simultaneous localization of MLL, AF4 and ENL genes in interphase nuclei by 3D-FISH: MLL translocation revisited. BMC Cancer 6:20–24. doi:10.1186/1471-2407-6-20

Habermann FA, Cremer M et al (2001) Arrangements of macro- and mini-chromosomes in chicken cells. Chromosome Res 9:569–584

Harewood L, Schütz F, Boyle S (2010) The effect of translocation-induced nuclear re-organization on gene expression. Genome Res 20(5):554–564. doi:10.1101/gr. 103622. 109

Holmquist GP (1992) Review article: chromosome bands, their chromatin flavors, and their functional features. Am J Hum Genet 51:17–37

Jakob B, Splinter J, Durante M, Taucher-Scholz G (2009) Live cell microscopy analysis of radiation-induced DNA double-strand break motion. Proc Natl Acad Sci USA 106:3172–3177

Jakob B, Splinter J et al (2011) DNA double-strand breaks in heterochromatin elicit fast repair protein recruitment, histone H2AX phosphorylation and relocation to euchromatin. Nucleic Acids Res 39(15):6489–6499. doi:10.1093/nar/gkr230

Jeppesen P (1997) Histone acetylation: a possible mechanism for the inheritance of cell memory at mitosis. Bioessays 19(1):67–74

Jeppesen P, Turner BM (1993) The inactive X chromosome in female mammals is distinguished by a lack of histone H4 acetylation, a cytogenetic marker for gene expression. Cell 74:281–289

Johannes C, Horstmann M et al (2004) Chromosome intrachanges and interchanges detected by multicolour banding in lymphocytes: searching for clastogen signatures in the human genome. Radiat Res 161:540–548

Karagiannis TC, Kn H, El-Osta A (2007) Disparity of histone deacetylase inhibition on repair of radiation-induced DNA damage on euchromatin and constitutive heterochromatin compartments. Oncogene 26:3963–3971

Korenberg JR, Rykowski MC (1988) Human genome organization: Alu, Lines and the molecular structure of metaphase chromosome bands. Cell 53:391–400

Küpper K, Köbl A et al (2007) Radial chromatin positioning is shaped by local gene density, not by gene expression. Chromosoma (Berl) 116:285–306

Louis SF, Vermolen BJ et al (2005) c-Myc induces chromosomal rearrangements through telomere and chromosome remodeling in the interphase nucleus. Proc Natl Acad Sci USA 102:9613–9618

Martínez-López W, Porro V et al (2000) Interchromosomal distribution of gamma ray-induced chromatid aberrations in Chinese hamster ovary cells. Genet Mol Biol 23:1071–1076

Martínez-López W, Folle GA, Jeppesen P, Obe G (2001) Chromosome regions enriched in hyperacetylated histone H4 are preferred sites for endonuclease- and radiation-induced breakpoints. Chromosome Res 9:69–75

Miller OJ, Therman E (2001) Human chromosomes. Springer, New York

Obe G, Durante M (2010) DNA double strand breaks and chromosomal aberrations. Cytogenet Genome Res 128:8–16

Obe G, Pffeifer P et al (2002) Chromosome aberrations: formation, identification and distribution. Mutat Res 504:17–36

O'Keefe RT, Henderson SC, Spector DL (1992) Dynamic organization of DNA replication in mammalian cell nuclei: spatially and temporally defined replication of chromosome-specific alpha satellite DNA sequences. J Cell Biol 116:1095–1110

Palitti F (1993) Mechanisms of induction of chromosomal aberrations by inhibitors of DNA topoisomerases. Environ Mol Mutagen 22(4):275–277

Parada LA, McQueen PG, Misteli T (2004) Tissue-specific spatial organization of genomes. Genome Biol 5:R44

Podhorecka M, Skladanoski K, Bozko P (2010) H2AX phosphorylation: its role in DNA damage response and cancer therapy. J Nucleic Acids 6:673–681. doi:10.4061/2011/920161

Postberg J, Lipps HJ, Cremer T (2010) Evolutionary origin of the cell nucleus and its functional architecture. Essays Biochem 48:1–24

Rogakou EP, Pilch PR et al (1998) DNA double-stranded breaks induce phosphorylation of histone H2AX on serine 139. J Biol Chem 273(10):5858–5868

Roix JJ, McQueen PG et al (2003) Spatial proximity of translocation-prone loci in human lymphocytes. Nat Genet 34:287–291

Rouquette J, Genoud C et al (2009) Revealing the high-resolution three-dimensional network of chromatin and interchromatin space: a novel electron-microscopic approach to reconstructing nuclear architecture. Chromosome Res 17:801–810

Sadoni N, Langer S et al (1999) Nuclear organization of mammalian genomes: polar chromosome territories build up functionally distinct higher order compartments. J Cell Biol 146(6):1211–1226

Sadoni N, Cardoso MC et al (2004) Stable chromosomal units determine the spatial and temporal organization of DNA replication. J Cell Sci 117:5353–5365

Simpson PJ, Savage JR (1995) Estimating the true frequency of X-ray-induced complex chromosome exchanges using fluorescence in situ hybridization. Int J Radiat Biol 67:37–45

Slijepcevic P, Natarajan AT (1994) Distribution of radiation-induced G_1 exchange and terminal deletion breakpoints in Chinese hamster chromosomes as detected by G-banding. Int J Radiat Biol 66:747–755

Solovei I, Kreysing M et al (2009) Nuclear architecture of rod receptor cells adapt to vision in mammalian evolution. Cell 137:356–368

Tanabe H, Habermann FA, Solovei I, Cremer M, Cremer T (2002) Non-random radial arrangements of interphase chromosome territories: evolutionary considerations and functional implications. Mutat Res 504:37–45

Weckerle AB, Santra M et al (2011) *CBFB* and *MYHII* in inv(16)(p13q22) of acute myeloid leukaemia displaying close spatial proximity in interphase nuclei of human hematopoietic stem cells. Genes Chromosomes Cancer 50:746–755

Xiao Y, Natarajan AT (1999) Non-proportional involvement of Chinese hamster chromosomes 3, 4, 8 and 9 in X-ray-induced chromosomal aberrations. Int J Radiat Biol 75:943–951

Chapter 4
Interphase Chromosomes of the Human Brain: The Biological and Clinical Meaning of Neural Aneuploidy

Ivan Y. Iourov, Svetlana G. Vorsanova, and Yuri B. Yurov

Abstract The human brain is generally assumed to be populated by cells that share identical genomes or diploid chromosome sets. However, interphase molecular cytogenetics has shown variable mosaic aneuploidy to be a new feature of brain cells. Interphase FISH analysis has estimated the amount of aneuploid cells as approximately 10 % (about 100 billion cells) in more than a trillion postmitotic neuronal and glial cells in the normal adult human brain. Paradoxically, aneuploidy appears to feature the mammalian brain despite representing a devastating condition in humans. Furthermore, neural aneuploidy rates vary during ontogeny. Aneuploidy rates are dramatically increased in early brain development, but decrease significantly in the postnatal period. Additionally, acquired aneuploidy affecting the brain is shown to be associated with neurodevelopmental and neurodegenerative disorders (i.e., autism, schizophrenia, ataxia-telangiectasia, Alzheimer's disease). Furthermore, interphase molecular cytogenetics allows for the analysis of genome organization at the chromosomal level in brain cells, which is, unfortunately, beyond the scope of current neuroscience and genome research. Nonetheless, a number of pilot reports have determined analyzing interphase chromosome spatial organization in neuronal nuclei to be promising for genetics/genomics and cell biology of the human brain.

I.Y. Iourov (✉)
Mental Health Research Center, Russian Academy of Medical Sciences,
Moscow 119152, Russia

Institute of Pediatrics and Children Surgery, Ministry of Health, Moscow, Russia
e-mail: ivan.iourov@gmail.com

S.G. Vorsanova • Y.B. Yurov (✉)
Mental Health Research Center, Russian Academy of Medical Sciences,
Moscow 119152, Russia

Institute of Pediatrics and Children Surgery, Ministry of Health, Moscow, Russia

Moscow City University of Psychology and Education, Moscow, Russia
e-mail: y_yurov@yahoo.com

Y.B. Yurov et al. (eds.), *Human Interphase Chromosomes: Biomedical Aspects*,
DOI 10.1007/978-1-4614-6558-4_4, © Springer Science+Business Media, LLC 2013

Introduction

The genomic landscape of the normal and diseased human brain had remained largely obscure until molecular cytogenetic or cytogenomic methods (i.e., fluorescence in situ hybridization, or FISH) for visualizing interphase chromosomes in nondividing neural cells became available (Vorsanova et al. 2010c). For several decades, indirect evaluations of neural chromosomes have resulted in confusion whether the human brain is populated by polyploid or diploid/euploid (normal) brain cells (Iourov et al. 2006c; Kingsbury et al. 2006; Mosch et al. 2007; Arendt et al. 2009). The dilemma has been resolved by interphase molecular cytogenetic studies, which have directly addressed genomic content of neural cells and have established that the overwhelming majority of cells populating the human brain are euploid.

Historically, the first attempt to evaluate chromosome numbers in the human brain was performed by Prof. van der Ploeg's group (Arnoldus et al. 1989, 1991, 1992). Their idea was referred to use interphase cytogenetics for studying genetic changes in brain tumors. Interphase nuclei isolated from unaffected brain tissues were analyzed as well. In the normal brain, they found relatively high levels of trisomy (mean rate, ~2 % per individual chromosome) (Arnoldus et al. 1989). Unfortunately, these data have not been appreciated by geneticists and neuroscientists. On the other hand, it has been repeatedly shown that almost all human somatic and germline tissues can contain a detectable amount of chromosomally abnormal cells as a result of sporadic (spontaneous) genome instability (Iourov et al. 2006a, 2008a, b; Hulten et al. 2008, 2010, 2013). Brain tissues are not an exception. Thus, it is hard to disagree with the idea that "aneuploidy is a necessary evil in human life" (Weier et al. 2010). First, aneuploidy in germline cells leads to the most common type of genetic pathology, termed "chromosomal diseases." Second, aneuploidy in somatic cells is involved in cancer pathogenesis (Duesberg et al. 2005). Finally, intrinsic aneuploidy rates in the human brain and its biological significance remain a matter of discussion (Iourov et al. 2010, 2012). Patterns of cellular variability and complexity in the central nervous system (Muotri and Gage 2006), integration of genetically abnormal neural cells into brain circuitry, and neuron–glia interactions (Kingsbury et al. 2005) allow us to speculate that neural aneuploidy plays a role in normal and pathogenic genome heterogeneity that is surely underestimated (Iourov et al. 2006c, d; Kingsbury et al. 2006).

In March 2005, three papers reevaluating aneuploidy in the normal human brain by interphase molecular cytogenetics were published. Professor Chun's group has focused on chromosome 21 aneuploidy in neural cells of the adult human brain. Surprisingly, they found chromosome 21 aneuploidy in about 4 % (40 billion?) of cells among approximately 1 trillion nonneuronal cells and postmitotic neurons in the human brain (Rehen et al. 2005). In comparison, human interphase lymphocytes show chromosome 21 aneuploidy rates in ~0.6 %. This study was unable to estimate the overall aneuploidy rates in the human brain as only one chromosome was analyzed. Nonetheless, it allowed speculation that all human beings are "low-level chromosome 21 trisomics" (or affected by mosaic trisomy 21/Down syndrome). Two other papers have evaluated chromosome complements in the developing and adult human brain by a quantitative FISH (QFISH) analysis (Iourov et al. 2005) and interphase FISH with a set of chromosome enumeration DNA probes specific to 13 chromosomes: 1, 7, 8, 9, 13, and 21; 14 and 22; 15, 16, 18, X, and Y (Yurov et al. 2005). Increased aneuploidy rates were found in cultured embryonic brain tissues as to the adult brain (1.3–7.0 % per individual chromosome, in contrast to 0.6–3.0 % in uncultured fetal brain cells and 0.1–0.8 % in postmortem adult brain cells, respectively). The overall aneuploidy incidence in the normal adult human brain was, therefore, estimated as nearly 10 %. These data have given rise to a hypothesis suggesting aneuploidy affects up to 100 billion of a trillion neuronal and nonneuronal cells populating the normal human brain.

The pilot neurocytogenetic studies have revealed significant aneuploid cell populations in the developing and adult human brain. Furthermore, aneuploidy affecting a larger amount of brain cells was found to be involved in pathogenesis of psychiatric and neurological (neurodegenerative) diseases (Yurov et al. 2001, 2007a, 2008; Iourov et al. 2006a, 2009a, b; Mosch et al. 2007; Boeras et al. 2008; Westra et al. 2008; Arendt et al. 2009, 2010; Granic et al. 2010). In addition, there is evidence that aneuploidy can be involved in normal and pathological brain aging (Iourov et al. 2008a; Yurov et al. 2009b; Granic et al. 2010; Faggioli et al. 2011; Fischer et al. 2012). Taken together, these observations have given rise to new directions in biomedical research—molecular neurocytogenetics and cytogenomics of brain diseases (Iourov et al. 2006c, 2008b).

Here, we consider current hypotheses concerning brain-specific genome variability, which probably plays a role in the etiology and pathogenesis of neuropsychiatric diseases. Additionally, we have tried to refer to all available neurocytogenetic studies covering the field of molecular neurocytogenetics that were published in peer-reviewed scientific journals during the past 10–12 years as well as reviews highlighting attractive hypotheses based on molecular cytogenetic and genomic data (Iourov et al. 2006c, d, 2008b, 2010, 2012; Kingsbury et al. 2006; Arendt 2012; Arendt et al. 2010; Zekanowski and Wojda 2009; Astolfi et al. 2010).We speculate that testing hypotheses concerning chromosome, genome, and epigenome variations in brain cells can be used for creating a unified theory considering the biological and clinical meaning of neural genome instability during ontogeny. The theory should provide for a coherent explanation of the role that somatic genome instability plays in the pathogenesis of genetically and etiologically heterogeneous brain diseases (autism, schizophrenia, ataxia-telangiectasia, and Alzheimer's disease) and brain aging.

Aneuploidy in the Developing Human Brain

The complexity and variability of the human brain are generated during the early prenatal development and are strongly determined by genomic content of neural progenitor cells (Muotri and Gage 2006). At early ontogeny, the murine developing brain possesses approximately 30 % of aneuploid cells (Rehen et al. 2001). Because the frequency of aneuploid conceptions (meiotic plus mitotic aneuploidy) usually differs significantly between species, one can question whether brain-specific aneuploidy in mice can model the phenomenon in humans (Iourov et al. 2006c; Hassold et al. 2007; Dierssen et al. 2009). However, this is not the case of the developing human brain. Molecular cytogenetic study of organotypic human neuronal cell cultures using interphase FISH with probes specific for chromosomes 1, 13/21, 18, X, and Y has found aneuploidy frequency to vary between 0.7 and 3 % per chromosome and to achieve 28 % in terms of the entire genome (Yurov et al. 2005). The elaboration of high-resolution molecular cytogenetic techniques, providing for visualization of interphase chromosomes at all stages of the cell cycle and at molecular resolutions, such as QFISH and interphase chromosome-specific multicolor banding (ICS-MCB), allowed us to be more accurate in estimating intercellular genomic variations at the chromosomal level in the developing human brain (Fig. 4.1) (Iourov et al. 2005, 2006a, 2007a; Vorsanova et al. 2010c). To address genomic variation during early development in more detail, aneuploidy and polyploidy were monitored in human fetuses (8–15 weeks of gestation). The developing human brain was found to have a mosaic nature, being composed of euploid and aneuploid neural cells. By studying more than 600,000 neural cells, the average aneuploidy frequency was estimated as 1.25–1.45 % per chromosome. The overall percentage of aneuploidy tended to approach 30–35 %. Tetraploidy affected about 0.04 % of embryonic neuronal cells (Yurov et al. 2007a). In total, these data provide evidence for aneuploidization in the developing brain to be evolutionarily conserved in mammals (Rehen et al. 2001; Yurov et al. 2005, 2007a; Iourov et al. 2006c). However, a unique feature of the developing human brain in terms of intercellular chromosomal/genomic variation was discovered: chromosome-specific aneuploidy is confined to the developing human brain (chromosome-specific low-level mosaic aneuploidy is exclusively confined to neural cell populations without affecting other fetal tissues) (Yurov et al. 2007a). It is to note that this is the only available report on aneuploidy mosaicism limited to an embryonic (not extraembryonic!) tissue.

Interestingly, the amount of aneuploid cells determined in the developing human brain (30–35 %) was found to approach the amount of cells cleared by programmed cell death (30–50 %) throughout human prenatal development (Muotri and Gage 2006; Yurov et al. 2007a). Therefore, considering the pathogenic effect of aneuploidy on cellular physiology (Dierssen et al. 2009), aneuploidization in the developing human brain was hypothesized to be a mechanism for neural cell number regulation by clearance of genetically abnormal and aneuploid cells either through apoptosis or through a cascade of mitotic catastrophes (Iourov et al. 2006c, d).

Fig. 4.1 (continued) of a chromosome (from *left* to *right*): chromosome 9 (**d**), chromosome 16 (**e**), and chromosome 18 (**f**). (**g**) Interphase QFISH: (1) a nucleus with two signals for chromosomes 18 (relative intensities: 2,058 and 1,772 pixels), (2) a nucleus with one paired signal mimics monosomy of chromosome 18 (relative intensity: 4,012 pixels), (3) a nucleus with two signals for chromosome 15 (relative intensities: 1,562 and 1,622 pixels), (4) a nucleus with one signal showing monosomy of chromosome 15 (relative intensity: 1,678 pixels) (From Yurov et al. 2007a. An open-access article distributed under the terms of the Creative Commons Attribution License)

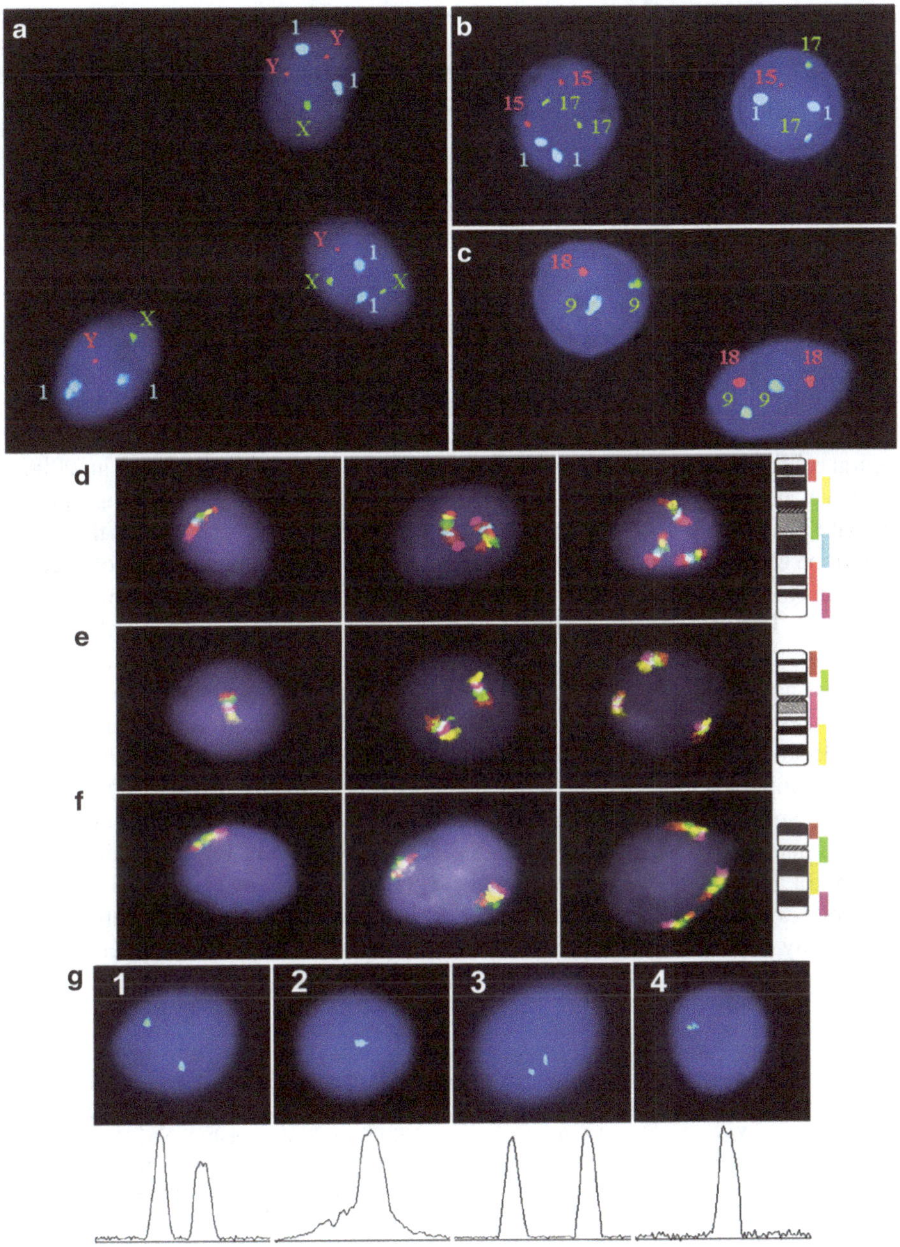

Fig. 4.1 Molecular cytogenetic analysis of aneuploidy in the fetal human brain. (**a–c**) Interphase fluorescence in situ hybridization (FISH) with chromosome-enumeration DNA probes: two nuclei characterized by additional chromosomes Y and X and a normal nucleus (**a**); a nucleus with monosomy of chromosome 15 and a normal nucleus (**b**); and a nucleus with monosomy of chromosome 18 and a normal nucleus (**c**). (**d–g**) Interphase chromosome-specific multicolor banding (MCB): nuclei with monosomy, disomy, trisomy, and G-banding ideograms with MCB color-code labeling

Aneuploidy is the most common type of mosaic chromosome instability (CIN) associated with the malignization process (Li et al. 2009; Weaver and Cleveland 2009). It was hypothesized that developmental instability of the genome confined to the brain cell populations has the potential to cause childhood brain cancer—the second most common childhood cancer after leukemia (Iourov et al. 2009c).

Aneuploidy in the Normal Adult Human Brain

The first analyses performed by single- and multiprobe FISH with chromosome-enumeration DNA probes have demonstrated neural aneuploidy rates per individual chromosome to vary in a wide range between 0 and 4 % or even more (Rehen et al. 2005; Yurov et al. 2005). Aneuploidy estimations indicate approximately 10 % of neural cells to be aneuploid in the adult brain (Yurov et al. 2005; Iourov et al. 2006a; Mosch et al. 2007; Westra et al. 2008). Although FISH is the technique most applied for interphase molecular cytogenetic analyses (Vorsanova et al. 2010c), there is a limitation of the classical interphase FISH protocols referred to the study of specific genomic loci without an integral view of the whole chromosome (Iourov et al. 2006b, d). Taking into account that neural CIN in the developing mammalian brain manifests almost exclusively as aneuploidy (Rehen et al. 2001; Yurov et al. 2007a), there has not been an empirical background for suggesting additional chromosomal imbalances in the unaffected human brain. Nonetheless, a need for further analyses by molecular cytogenetic techniques providing for visualization of the whole chromosome appeared to exist. The latter was solved by ICS-MCB (Fig. 4.2), the only available approach offering such opportunities that allowed identifying more precise rates of aneuploidy per individual chromosomes in the adult brain, but the overall amount of aneuploid cells still remained at about 10 % (Iourov et al. 2006a, 2007a). This rate was also confirmed in control brain samples used for the evaluation of aneuploidy in human brain diseases (i.e., schizophrenia, ataxia telangiectasia, and

Fig. 4.2 (continued) spond to 18p11.2Yq12.2. SO (Spectrum Orange) signals Y 18p11.2Yp11.3. TR (Texas Red) signals Y 18q22Yq23. Cy5 signals Y 18q11.2Yq21.3. (**e**) FISH with MCB probe for chromosome X.R110 signals correspond to Xp21.3Yp22.3 and Xq25Yq28. SO (Spectrum Orange) signals Y Xp11.22Yp22.1 and Xq25Yq28. TR (Texas Red) signals Y Xq12Yq21.1. Cy5 signals Y Xq21.1Yq26. DEAC signals Y Xp11.3Yq13. Note the upper chromosome X appears as a *white condensed spot* (merged image). Because facultative heterochromatin, a feature of X chromosome inactivation, should appear as a highly condensed structure, the upper X chromosome was assumed to be the inactivated one (Xi) in contrast to the active X chromosome (Xa) appearing as a lightly diffused structure. (**f**) Example of a trisomic nucleus (trisomy of chromosome 9): *left side* Y black-and-white picture of DAPI-counterstained nucleus, *right side* Y merged MCB true color picture showing the presence of three chromosomes 9 in this nucleus. (**g**) Example of a monosomic nucleus (monosomy of chromosome 18): *left side* Y black-and-white picture of DAPI-counterstained nucleus, *right side* Y merged MCB true color picture showing the presence of one chromosome 18 in this nucleus (From Iourov et al. 2006a. Reproduced with permission of Elsevier BV in the format reuse in a book/textbook via Copyright Clearance Center)

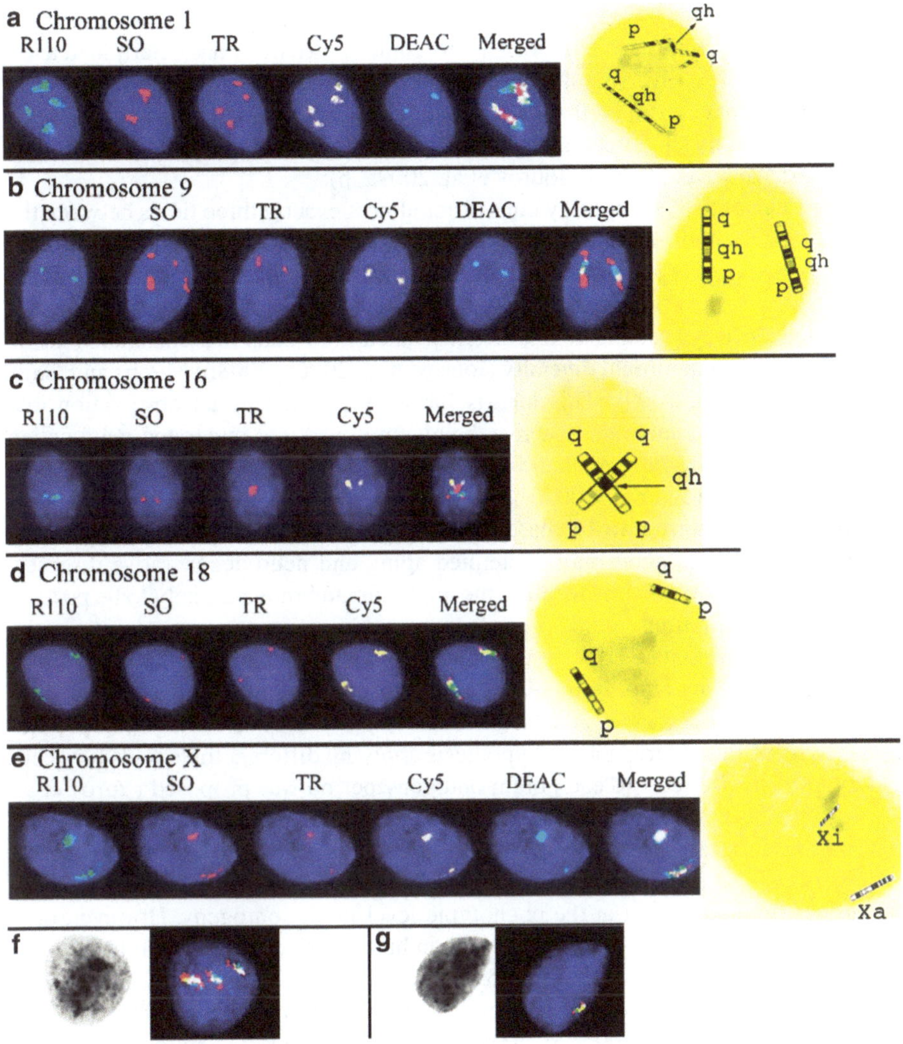

Fig. 4.2 FISH using MCB probes on interphase nuclei of the human brain. (**a**) FISH with MCB probe for chromosome 1. R110 signals correspond to 1p32.3Yp36.3 and 1q32Yq43. SO (*Spectrum Orange*) signals Y 1p13Yq21, including constitutive heterochromatin (1qh). TR (*Texas Red*) signals Y 1p31.1Yp33 and 1q21.3Yq31. Cy5 signals Y 1p13.1Yp22.3 and 1q32Yq43. DEAC signals Y 1q21.3Yq31. Note the upper chromosome 1 is folded around 1qh and bent in the proximal part of the q-arm. (**b**) FISH with MCB probe for chromosome 9. R110 signals correspond to 9p13Yq13 including constitutive heterochromatin (9qh). SO (Spectrum Orange) signals Y 9p21Yp24 and 9q32Yq34. TR (Texas Red) signals Y 9q22.2Yq34.1. Cy5 signals Y 9p13Yp23. DEAC signals Y 9q13Yq22.2. (**c**) FISH with MCB probe for chromosome 16. R110 signals correspond to 16p11.1Yp13.1 SO (Spectrum Orange) signals Y 16p13.3Yp21. TR (Texas Red) signals Y 16q11.1Yq21 including constitutive heterochromatin (16qh). Cy5 signals Y 16q21Yq24. Note the single Texas Red signal instead of two; this implies that 16qh regions of two homologous chromosomes 16 are overlapped. Therefore, somatic pairing of two homologous chromosomes 16 by 16qh region should be suspected. (**d**) FISH with MCB probe for chromosome 18. R110 signals corre

Alzheimer's disease) (Yurov et al. 2008; Iourov et al. 2009a, b). Additionally, a recently proposed approach to define "DNA content variation" has determined average genome content diversification between neuronal cells as ~250 Mb (Westra et al. 2010). These data accord well with observations on aneuploidy in the adult human brain performed by single-cell interphase molecular cytogenetic approaches (Mosch et al. 2007; Westra et al. 2008; Iourov et al. 2009a, b).

As one can notice, aneuploidy rates differ almost exactly three times between the developing and adult human brain. Therefore, suggestions about neural aneuploid clearance throughout prenatal development appear to be consistent with data on the postnatal brain. Nevertheless, the biological role of aneuploidy in the adult human brain remains to be established. Currently, aneuploid cells are considered to be involved in human neuronal diversity (Iourov et al. 2006c, 2008b; Muotri and Gage 2006; Arendt et al. 2009). This idea is further supported by an observation that aneuploid cells are functionally active, being employed into integrated mammalian brain circuitry (Kingsbury et al. 2005). Moreover, aneuploidy is probably involved in brain aging (Yurov et al. 2009b, 2010; Faggioli et al. 2011, 2012), inasmuch as aneuploidy rates appear to increase during postnatal ontogeny stages and aneuploidy is involved in abnormal/accelerated aging and neurodegenerative diseases. Because the majority of cells forming the adult human brain are likely to be postmitotic, these observations seem to produce a paradox. Somatic aneuploidy results largely from abnormal cell divisions during neurogenesis in the early brain development. Therefore, aneuploidy increase in late ontogeny may be only explained by widespread adult neurogenesis, which is unable to produce such a large cell populations. To solve this discrepancy, a hypothesis applying different thresholds for aneuploidy levels and effects to each brain ontogeny period was proposed (Yurov et al. 2009b). The latter suggests constitutional and acquired aneuploidy to alter cooperatively the homeostasis of neural cells (neurons and glia) during ontogeny, to generate senescent cellular phenotypes (probably, promoting cell death), but these processes begin to become apparent at the phenotypic level in late ontogeny. However, only direct experimental aneuploidy monitoring in human brain aging would help to test this hypothesis and to solve the paradox.

The effect of aneuploidy on human cell populations is known to be extremely devastating (Iourov et al. 2006c, d, 2008b; Hassold et al. 2007; Dierssen et al. 2009). Thus, one can assume brain aneuploidization to be pathogenic in contrast to hypotheses proposing a role of aneuploidy in neural diversity. To define benign sporadic aneuploidy in the adult human central nervous system, it is necessary to compare the amount of aneuploid cells between the normal and diseased human brain.

Aneuploidy in the Diseased Human Brain

The diseases associated with brain dysfunction and aneuploidy are chromosomal aneuploidy syndromes: autosomal and gonosomal trisomies, an additional chromosome X in males, and chromosome X monosomy in females (Iourov et al. 2006c,

2008b; Hassold et al. 2007; Dierssen et al. 2009). Direct molecular cytogenetic evaluations of the brain are exclusive in chromosomal aneuploidy syndromes. Nevertheless, these pathological conditions were used for models and hypotheses of brain diseases in the widest sense and their probable association with mosaic aneuploidy in the brain (Yurov et al. 2001; Iourov et al. 2006c, 2008b). As a result, autism, schizophrenia, ataxia-telangiectasia, and Alzheimer's disease have been directly assessed by a series of molecular neurocytogenetic studies (Yurov et al. 2001, 2008; Iourov et al. 2009a, b; Mosch et al. 2007; Yang and Herrup 2007; Boeras et al. 2008; Westra et al. 2009; Arendt et al. 2009, 2010). In addition, numerous brain diseases are hypothesized to be associated with brain-specific aneuploidy or CIN. Table 4.1 provides an overview of the latest molecular neurocytogenetic achievements in brain research.

Autism

Autism is an umbrella term for a number of neurodevelopmental disorders characterized by etiological and genetic heterogeneity including more than 100 genetic and genomic diseases (Betancur 2011). Autism is frequently associated with chromosomal imbalances (Castermans et al. 2004; Xu et al. 2004). Using cytogenetic and molecular cytogenetic techniques, constitutional chromosomal abnormalities are found in about 5–7 % of autism cases (Xu et al. 2004; Vorsanova et al. 2007, 2010a, b). The contribution of mosaic aneuploidy to autism pathogenesis is estimated as 16 %, probably representing the most common molecular cytogenetic finding in children with unexplained autism (Fig. 4.3). It is to be noted that 10 % of males with unexplained autism exhibited low-level 47,XXY/46,XX mosaicism (Yurov et al. 2007b). This finding was used for a hypothesis suggesting mosaic X chromosome aneuploidy to be involved in male predisposition to autistic spectrum disorders (Iourov et al. 2008c). Finally, a recent study has shown mosaic aneuploidy and CIN to segregate with mental diseases in autistic families (Vorsanova et al. 2010b). Interestingly, Rett syndrome, an X-linked autistic spectrum monogenic disease, associated with male prenatal lethality, has been found to occur in males who are 47,XXY/46,XY mosaics (Vorsanova et al. 1996, 2001). Additionally, mosaicism in Rett syndrome males was tissue specific and was confined to ectodermal tissues (Vorsanova et al. 2001). As the disease is primarily associated with neurodevelopmental abnormalities, it was assumed that the majority of (if not all) boys with Rett syndrome should have cells with additional chromosome X in the affected brain (Yurov et al. 2001; Iourov et al. 2006c, 2008a).

Molecular neurocytogenetic studies have revealed somatic genome instability or mosaic aneuploidy to increase in the developing central nervous system and appear to play a role in brain development. It was hypothesized that neuronal aneuploidy alters brain development and is involved in male predisposition to autism or related psychiatric conditions (Iourov et al. 2006a). To test this hypothesis we have attempted to estimate the incidence of mosaic aneuploidy in the autistic brain tissue

Table 4.1 Chromosome/genome instabilities in brain disorders

Disease	Brief overview of study design	Main outcome	Key references
Direct evaluations of the human brain			
Schizophrenia	Multiprobe FISH	Two cases of six exhibited low-level mosaic trisomy of chromosomes 18 and X	Yurov et al. (2001)
	Multiprobe FISH, QFISH, ICS-MCB	Two cases of 12 exhibited low-level mosaic aneuploidy (monosomy and trisomy) of chromosome 1 (Fig. 4.1c); sporadic chromosome 1-specific aneuploidy was increased in the diseased brain	Yurov et al. (2008)
Alzheimer's disease	Slide-based cytometry followed by single-probe FISH and chromogenic in situ hybridization	Aneuploidy confined to brain tissues is increased in preclinical stages of the disease and is involved in abnormal neuronal cell death	Mosch et al. (2007) and Arendt et al. (2010)
	Single-probe FISH	The frequency of tetraploid (tetrasomic?) cells is increased in the diseased brain	Yang and Herrup (2007)*
	Multiprobe FISH, QFISH, ICS-MCB	Chromosome 21-specific aneuploidy (Fig. 4.1d) is dramatically increased in the diseased brain and is associated with neurodegeneration	Iourov et al. (2009b)
	Two-probe FISH	Lack of tetraploidy increase in the diseased brain	Westra et al. (2009)
Ataxia-telangiectasia	Multiprobe FISH, QFISH, ICS-MCB	Aneuploidy of all chromosomes observed in all brain areas (Fig. 4.1e) and breakages of chromosome 14 producing additional rearranged chromosomes (Fig. 4.1f) are confined to the degenerated cerebellum affecting ~40 % of cells	Iourov et al. (2009a, b)
Indirect evaluations			
Alzheimer's disease	Extensive set of cytogenetic and molecular cytogenetic techniques**	Aneuploidy is rarely observed in blood lymphocytes, but is reported to affect skin fibroblasts	Potter (2008)* and Iourov et al. (2010)*
	Single-probe FISH on transfected human presenelin 1-mutated cells	Acquired chromosome missegregation causing aneuploidy associated with mutated presenelin 1	Boeras et al. (2008)
	Single-probe FISH on transgenic mice and transfected human cells	Amyloid precursor protein gene (*APP*) induce chromosome missegregation and aneuploidy	Granic et al. (2010)

Ataxia-telangiectasia	Two-probe FISH on brain cells of ataxia-telangiectasia murine model	A 38 % increase of sex chromosome aneuploidy in the murine brain (*Atm*$^{-/-}$ mice)	McConnell et al. (2004)
Autism	Multiprobe FISH, QFISH**	Chromosomal mosaicism and heteromorphism is highly prevalent among affected children	Vorsanova et al. (2007)
	Multiprobe FISH, QFISH**	16 % of autistic boys exhibit low-level mosaic aneuploidy; the most prevalent condition is mosaic 47,XXY/46,XY	Yurov et al. (2007b)
	Multiprobe FISH, QFISH**	Chromosomal mosaicism and heteromorphism cosegregates with autism and/or other mental diseases in affected families	Vorsanova et al. (2010b)
Chromosomal syndromes	Extensive set of cytogenetic and molecular cytogenetic techniques**	From 1 to 60 % of cases (depending on chromosome) of chromosomal aneuploidy syndromes are mosaic. In the remaining cases, aneuploidy is thought to affect all cells (including brain cells)	Iourov et al. (2006c*, d*, 2008a*, b*), Hassold et al. (2007)* and Dierssen et al. (2009)*
Mental retardation	Extensive set of cytogenetic and molecular cytogenetic techniques**	Up to 20–30 % of cases are associated with chromosome abnormalities; 3.5 % of cases exhibit chromosomal mosaicism	Iourov et al. (2006c*, d*, 2008a)*
Rett syndrome in males (X-linked dominant diseases)	Multiprobe FISH**	Boys with Rett syndrome are usually 47,XXY/46,XY mosaics to escape intrauterine death; mosaicism can be tissue specific	Vorsanova et al. (2001)
Schizophrenia	Extensive set of cytogenetic and molecular cytogenetic techniques**	Aneuploidy is observed in 1–4 % of cases; among them, there are mosaic and non-mosaic cases; the most frequent is sex chromosome aneuploidy	Iourov et al. (2006c*, 2008a*, b*), Yurov et al. (2008)*, and de Moraes et al. (2010)

*Reviewed by
**Blood lymphocytes/skin fibroblasts

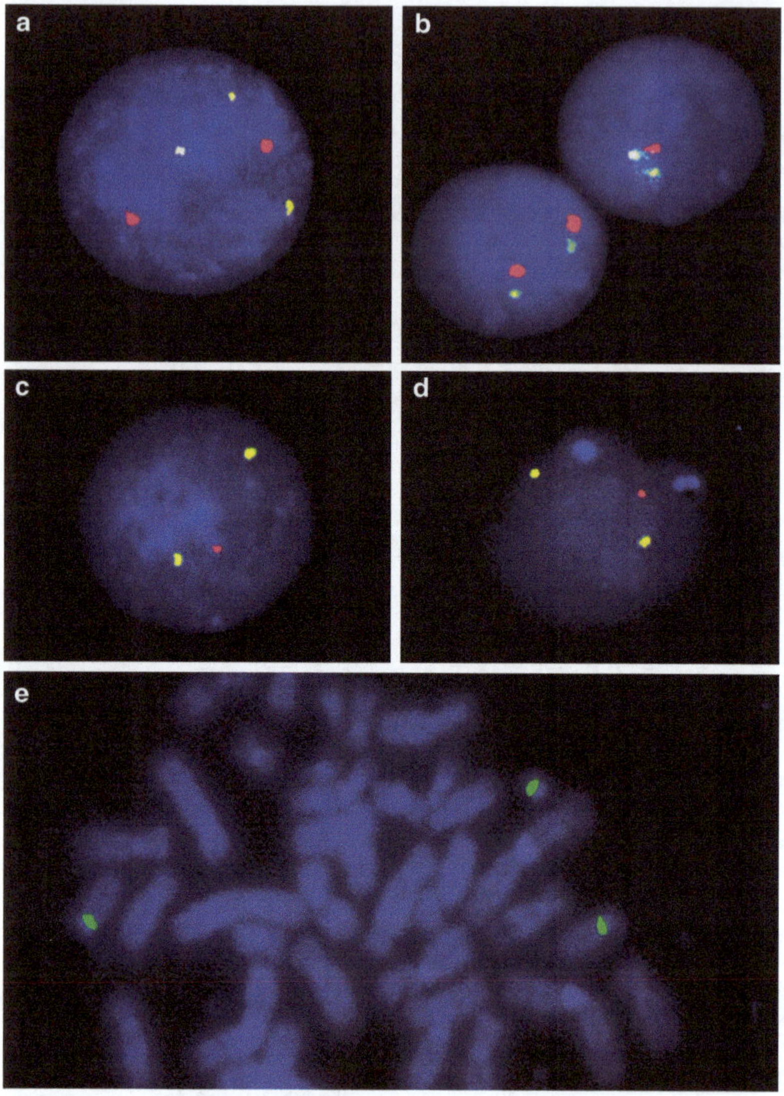

Fig. 4.3 FISH with chromosome-enumeration DNA probes in autism. (**a**) Nucleus characterized by trisomy 15 (three *green signals*) and two copies of chromosome 17 (two *red signals*). (**b**) A nucleus with monosomy 18 (one *red signal*) and a normal nucleus with disomy 18 (two *red signals*). Two chromosomes 9 are present in each nuclei (two *green signals*). (**c**) A nucleus with disomy X (two *green signals*) and one chromosome Y (one *red signal*). (**d**) A nucleus with disomy X (two *green signals*), two chromosomes 1 (two *light blue signals*), and one chromosome Y (one *red signal*). (**e**) Metaphase with additional chromosome der(15) and two normal chromosomes 15 (*green signals* at the centromeric regions of chromosomes 15) (From Yurov et al. 2007b, *Journal of Medical Genetics* by the British Medical Association. Reproduced with permission of BMJ Publishing Group in the format reuse in a book/monograph via Copyright Clearance Center)

using molecular cytogenetic techniques. Postmortem brain tissues of 12 patients with idiopathic autism, obtained from the NICHD Brain and Tissue Bank for Developmental Disorders at the University of Maryland, Baltimore, MD, USA, were analyzed using a chromosome-X-specific alphoid DNA probe (Yurov et al. 2011b, 2012). In this pilot interphase cytogenetic study, we observed statistically significant increase of chromosome X aneuploidy rates in the cerebral cortex and cerebellum in the male autistic brain as compared to control samples. Autistic spectrum disorders currently affect four times as many males as females. Mosaic chromosome X aneuploidy in the brain may help to explain the preponderance of autism among males in addition to specific alterations of the X-chromosome genes. We conclude that intercellular genomic variation manifesting as brain-specific low-level mosaic aneuploidy is one of the possible genetic factors likely contributing to autism neuropathology. This finding agrees with the hypothesis that increased developmental instability of the somatic genome could affect neuronal homeostasis and functions of the autistic brain, playing, therefore, a role in the pathogenesis of this common nervous system disease. These data form a firm basis for forthcoming systematic molecular neurocytogenetic studies of the autism brain.

Schizophrenia

In addition to autism, there are increasing lines of evidences linking genomic and epigenomic instability (GIN), including CIN, to schizophrenia (Smith et al. 2010). Schizophrenia was the first disease studied through direct molecular neurocytogenetic evaluation (Yurov et al. 2001). Analyzing six samples of the postmortem schizophrenia brain by multiprobe FISH has shown two individuals to be both affected by low-level mosaic trisomy of chromosomes 18 and X. These data were intriguing in the light of numerous studies of individuals suffering from schizophrenia by an extensive set of cytogenetic and molecular cytogenetic during the past 40 years, which have shown from 1 to 4 % of patients exhibit sex chromosome aneuploidy as well as single cases of partial monosomy/trisomy of autosomes (DeLisi et al. 1994, 2005; Iourov et al. 2006c, 2008a, b; Yurov et al. 2008; de Moraes et al. 2010). More detailed molecular-cytogenetic evaluation of a cohort of 12 patients by multiprobe FISH/QFISH and ICS-MCB has discovered two additional cases of low-level mosaic aneuploidy confined to the schizophrenia brain: monosomy and trisomy of chromosome 1 (Fig. 4.4). Moreover, chromosome 1-specific sporadic aneuploidy is increased in the brain samples among those schizophrenia patients (Yurov et al. 2008). It is to be noted that chromosome 1 aneuploidy is one of the most devastating numerical chromosome imbalances usually associated with early embryonic lethality (Vorsanova et al. 2005; Iourov et al. 2006c). However, affecting less than 4–5 % of cells and limited to brain tissue, chromosome 1 aneuploidy seems to produce tissue-specific pathology (Iourov et al. 2008a, b). These lines of evidences allow the hypothesis that mosaic aneuploidy in the human adult brain is a likely mechanism for psychotic disorders such as schizophrenia, at least in some cases.

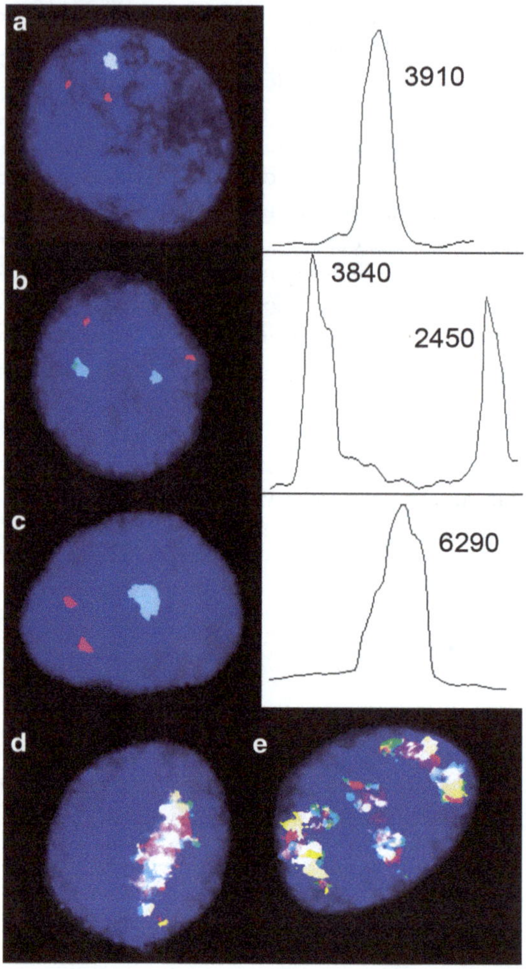

Fig. 4.4 Molecular cytogenetic analysis of aneuploidy in the postmortem schizophrenia brain. Interphase FISH with chromosome-enumeration DNA probes: a nucleus with monosomy involving chromosome 1 (one *white signal*, relative intensity: 3,910) and disomy X (two *red signals*) (**a**); a nucleus with disomy 1 (two *white signals*, relative intensities: 3,840 and 2,450) and disomy X (two *red signals*) (**b**); a nucleus with disomy 1 (one large *white signal* composed from two paired signals, relative intensity: 6,290) and disomy X (two *red signals*) (**c**). Interphase chromosome-specific MCB: nuclei with monosomy (**d**) and trisomy (**e**) involving chromosome 1 (From Yurov et al. 2008. Reproduced with permission of Elsevier BV in the format reuse in a book/textbook via Copyright Clearance Center)

Ataxia-Telangiectasia

Ataxia-telangiectasia (AT) is an autosomal recessive syndrome associated with CIN. This disease exhibits targeted cerebellar neurodegeneration, whereas other brain areas are paradoxically less affected (McKinnon 2004). To solve this paradox, a hypothesis suggesting CIN to affect selectively degenerating brain areas was proposed (Iourov et al. 2007b). The murine model ($Atm^{-/-}$ mouse) has demonstrated an appreciable increase of sex chromosome aneuploidy in the brain compared to unaffected mice (Table 4.1), but area-specific aneuploidy distribution has not been observed (McConnell et al. 2004). However, it is to be noted that $Atm^{-/-}$ mice do not demonstrate progressive cerebellar neurodegeneration, being poorly applicable for modeling ataxia-telangiectasia neuropathology.

Interphase cytogenetics using multiprobe FISH/QFISH and ICS-MCB have demonstrated a significant increase of aneuploidy in the ataxia-telangiectasia brain, achieving 20–50 % (Iourov et al. 2009b). Although dramatic neural aneuploidization was found to be a striking feature of this disease, the ataxia-telangiectasia paradox (Iourov et al. 2007b) was not completely solved. This lack led to an interphase chromosome study of different areas within the ataxia-telangiectasia brain by multiprobe FISH/QFISH and ICS-MCB followed by an in silico analysis. The cerebellum has shown a new CIN pattern distinct from that observed in the cerebrum. Apart from increased sporadic aneuploidy, chromosome-specific aneuploidy and nonrandom DNA double-strand breaks of chromosomes 14, 7, and, to a lesser extent, chromosome X, were discovered (Fig. 4.5). These breaks produced rearranged chromosomes in about 40 % of cerebellar cells, manifested essentially as der(14)(14pter->14q12:), and multiple aneuploidy involving rearranged chromosomes 14. The hotspots for targeted cerebellar neurodegeneration revealed by ICS-MCB and in silico analysis were mapped to 14q12, containing two candidate genes: *NOVA1* and *FOXG1B* (Fig. 4.6). It is known that Nova is a key brain-specific alternative splicing regulator in the vertebrate central nervous system. If a connection between impaired genome stability caused by *ATM* gene mutation and an aberrant process of genome regulation by NOVA1 does exist, it may provide elucidation of the pathogenic pathway of ATM-dependent neurodegeneration associated with aberrant splicing in cerebellar cells. The second prioritized gene (*FOXG1B*) encoding a transcriptional factor is known to regulate neurogenesis and is highly expressed in the fetal brain. Mutations in *FOXG1B* gene cause a clinical phenotype similar to Rett syndrome. Interestingly, the forkhead protein FoxG1 interacts with the methyl-CpG binding protein 2 (MeCP2, mutated in Rett syndrome) in mouse neurons. In differentiated neurons of the adult brain, *FOXG1B* promotes survival of postmitotic neurons, and its downregulation leads to neuronal cell death (Dastidar et al. 2012). One can propose that somatically acquired CIN and breakpoints in *FOXG1B* lead to its downregulation and promote neuronal death in the AT cerebellum. Thus, molecular neurocytogenetics provides a link between cerebellar dysfunction in neurodevelopmental and neurodegenerative disorders.

The speculations about GIN involvement in neurodegenerative and neurodevelopmental processes within the AT cerebellum define the ATM-directed selective increase of aneuploidy and chromosome-specific breaks to affect specific pathways of brain development and neuronal survival. Mosaic expression of GIN selectively in the cerebellum could help to explain the AT paradox, highlighted by McKinnon (2004). Identification of genes abnormally regulated in the AT brain will open new ways to explore cerebellar degeneration pathways and to develop targeted therapy in this, presently incurable, brain disorder (Yurov et al. 2009a).

Therefore, AT demonstrates that single-gene neurodegenerative diseases could be associated with chromosome-specific instability and aneuploidy confined to specific brain areas. In this instance, we have hypothesized that neurodegeneration and cancer has the same mechanism—genome and chromosome instabilities (Iourov et al. 2009a; Li et al. 2009; Weaver and Cleveland 2009). An additional implication

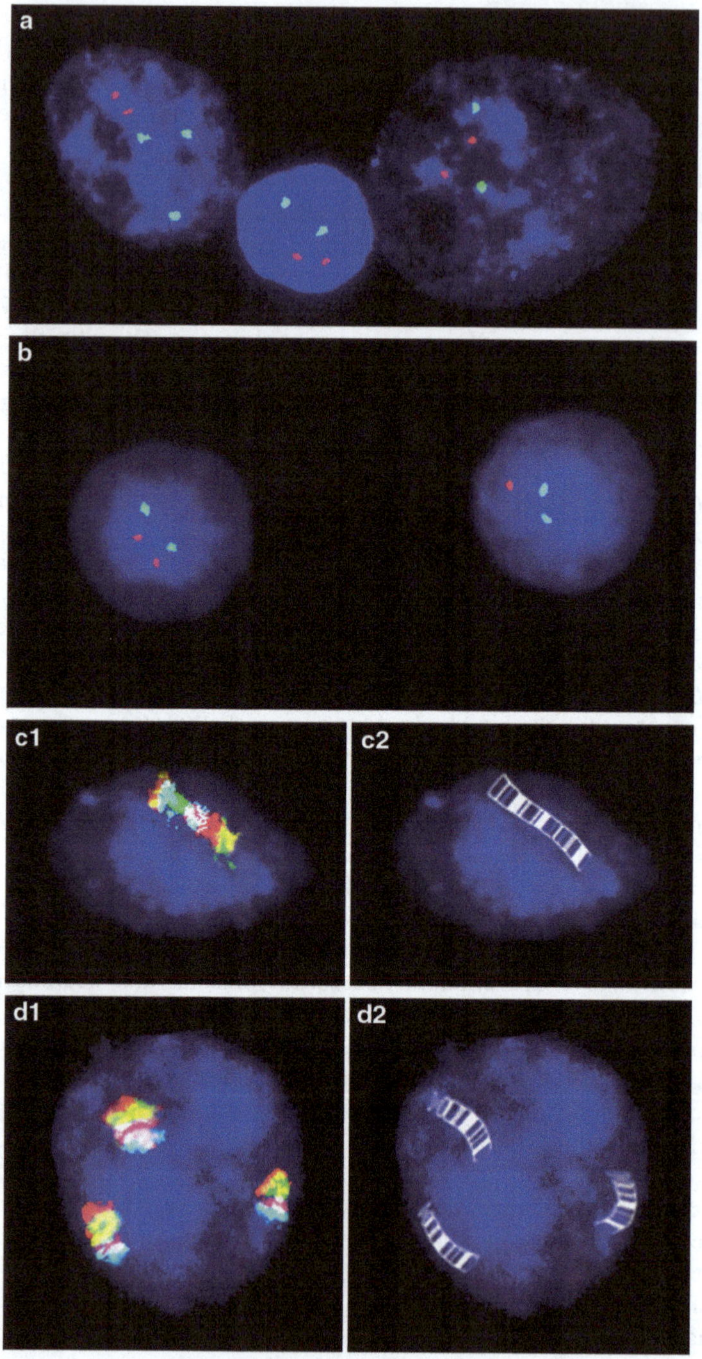

Fig. 4.5 Molecular cytogenetic analysis of aneuploidy in the cerebellum of the ataxia-telangiectasia (AT) brain by multiprobe FISH and ICS-MCB techniques. (**a**) True trisomy 7

of cerebellar neurodegeneration mechanism in ataxia-telangiectasia makes a basis for future successful strategies of therapeutic interventions by cell replacement therapy, which should be started immediately after birth (Yurov et al. 2009a).

Alzheimer's Disease

Alzheimer's disease (AD) was long thought to be associated with aneuploidy involving trisomy 21 (Heston and Mastri 1977; Potter 2008). It was known that individuals with Down's syndrome frequently develop AD-like neuropathology, and it was suggested that classical AD (genetic and late-onset sporadic forms) might be promoted by mosaic trisomy 21. More precisely, because of neuropathological parallels between AD and Down's syndrome, it has been hypothesized that individuals with AD should exhibit mosaic aneuploidy of chromosome 21 (Heston and Mastri 1977; Geller and Potter 1999; Potter 2008). Genetic mutations causing familial AD disrupt the cell cycle and lead to chromosome aneuploidy, including trisomy 21. However, until recently, no consensus has been obtained regarding the trisomy 21 hypothesis of AD pathogenesis (Potter 2008; Yurov et al. 2009b; Iourov et al. 2010).

Arendt and colleagues have shown that neurons with more-than-diploid DNA content are increased in preclinical AD stages and are selectively affected by cell death during disease progression (Arendt et al. 2010). Therefore, GIN or neuronal hyperploidy should be associated with decreased viability of neural cells in AD. Neuronal hyperploidy is, thereby, a direct molecular signature of cells prone to death in AD and indicates that a neuronal differentiation failure is a critical event in the AD pathogenetic cascade. Scoring a larger amount of neuronal cells by slide-based cytometry followed by single-probe FISH and chromogenic in situ hybridization, it was found that aneuploidy is likely to be increased in the AD brain (Mosch et al. 2007). Finally, direct analysis of the diseased brain using multiprobe FISH/QFISH and ICS-MCB has discovered chromosome 21-specific aneuploidy to increase dramatically (from 5- to 20 fold) in the AD cerebrum, and it was found to be involved in targeted neurodegeneration

Fig. 4.5 (continued) revealed by mFISH with chromosome 7-specific alphoid DNA probe (three *green signals*) in neuronal nucleus (*left*) in the cerebellum of the AT brain. Glial-like nucleus (*center*) and neuronal-like nucleus (*right*) with two *green signals*, indicating disomy 7. Chromosome X-specific alphoid DNA probe (*red signals*) indicates the presence of two copies of chromosome X in each nucleus. (**b**) Disomy X (nucleus in *left*, two *red signals*) and monosomy X (one *red signal*, nucleus in *right*) revealed by chromosome X-specific probe in the cerebellum of a woman with AT. Chromosome 7-specific alphoid DNA probe (*green signals*) indicates the presence of two copies of chromosome 7 in each nucleus. (**c1**) ICS-MCB with chromosome 7-specific MCB probe demonstrates monosomy 7 in neuronal nucleus of the AT brain. (**c2**) Scheme illustrates ideogram of chromosome 7 with G-banding in neuronal nucleus with monosomy 7. (**d1**) ICS-MCB with chromosome 14-specific MCB probe demonstrates trisomy 14 in neuronal nucleus of the AT brain. (**d2**) Scheme illustrates ideogram of chromosome 14 with G-banding in neuronal nucleus with trisomy 14 (From Yurov et al. 2009b. Reproduced with permission of Oxford University Press in the format reuse in a book/textbook via Copyright Clearance Center)

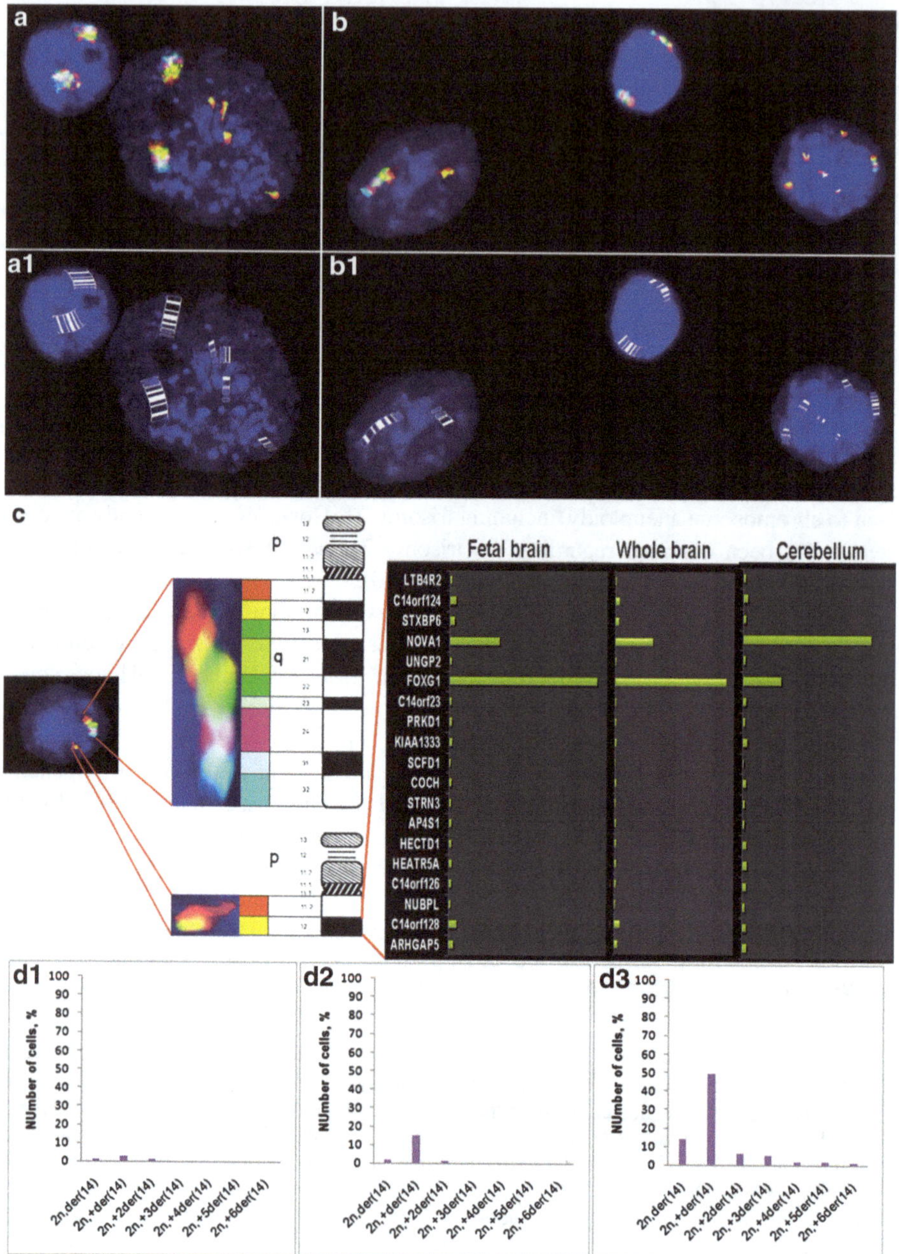

Fig. 4.6 Molecular cytogenetic analysis of chromosome 14 breaks in the cerebellum of the AT brain by ICS-MCB techniques. (**a**) FISH with chromosome 14-specific MCB probe demonstrates one neuronal nucleus (*left*) with two undamaged chromosome 14 (or disomy 14) and another nucleus with two undamaged chromosome 14 with additional four derivate chromosomes 14q12 (*right*). (**a1**) Scheme illustrates ideograms of undamaged and damaged chromosomes 14 with G-banding in the same nuclei as in (**a**). (**b**) FISH with chromosome 14-specific MCB probe demonstrates one

(Iourov et al. 2009b). Additionally, experimental and theoretical evaluations have shown that aneuploidy is probably involved in disease-causing selective neuronal cell death (Arendt et al. 2010). Thus, the hypothesis suggesting a common background in AD and Down's syndrome (Potter 2008) was confirmed. Moreover, mutated presenilin 1 and amyloid precursor protein gene cell lines (models of genetic defects associated with monogenic AD) were shown to exhibit high levels of aneuploidy (Table 4.1), suggesting these mutations promote aneuploidization (Boeras et al. 2008; Granic et al. 2010; Borysov et al. 2011). Therefore, chromosome 21 aneuploidy represents an integral component of the AD neurodegeneration pathogenic cascade (Yurov et al. 2009b; Iourov et al. 2010). However, aneuploidy in the AD brain demonstrates both chromosome 21 gain and loss, as well as affecting, in lesser instances, other chromosomes, including chromosome X (Fig. 4.7). These findings and studies of nonneuronal tissues indicate that not only trisomy 21 but another type of aneuploidy, or CIN, may be involved in the AD neurodegeneration pathway (Thomas and Fenesh 2008; Migliore et al. 2011; Spremo-Potrapevic et al. 2011; Taupin 2011). Thus, the hypothesis that AD is a mosaic of Down syndrome is attractive, but direct comparison of the pathogenic pathways associated with chromosome/genome instability in AD and Down's syndrome should to be performed with caution, requiring additional experimental proof (Potter et al. 2011).

A line of evidence concerning the high rates of polyploidy and abnormal DNA replication activity in the AD brain was provided. Because the overwhelming majority of cells in the human brain are considered to be postmitotic, it has been suggested that neurons enter the cell-cycle stage accompanied by chromosomal DNA replication but are unable to end the division (endomitosis or endoreplication). As a result, these neurons become tetraploid (Yang et al. 2001; Yang and Herrup 2007; Herrup and Yang 2007; Chen et al. 2010). Cell-cycle events including complete chromosomal DNA replication should ultimately result in generation of tetraploid cells. The empirical finding of tetraploid neurons at a higher frequency (to 4 %) in the AD hippocampus allowed the proposal that DNA replication precedes neuronal

Fig. 4.6 (continued) neuronal nucleus (*left*) with one undamaged chromosome 14 with additional der14q12; glial-like nucleus with disomy 14 (*center*); and another nucleus with one undamaged chromosome 14 with additional five derivate chromosomes 14q12 (*right*). (**a1**) Scheme illustrates ideograms of undamaged and damaged chromosomes 14 with G-banding in the same nuclei as in (**b1**). (**c**) *Left*: Neuronal nuclei with one undamaged chromosome 14 and one additional der14q12. *Center*: Two ideograms of undamaged chromosome 14 and one additional der14q12 with MCB labeling scheme. Chromosome der14q12 contains two labeled bands: q11.2 (*red*) and q12 (*yellow*). The majority of chromosomes der14 revealed in the diseased cerebellum (Fig. 4.5a,b) have the same MCB banding, indicating that DNA double-strand breaks occurred in the band 14q12 with the loss of the distal part of chromosome 14. *Right*: Levels of expression of 19 known genes mapped to the band 14q12 in the fetal human brain, in the whole human brain, and in the cerebellum, indicating that only two genes from chromosome 14q12 are highly expressed in the cerebellum: *NOVA1* and *FOXG1B*. (**d1**), (**d2**), (**d3**) The frequency of aneuploidy involving undamaged and damaged chromosome 14 in neural nuclei in the cerebellum of AT patients: patient UMB#1038, age 24 years (**d1**); patient UMB#1004, age 35 years (**d2**); patient UMB#878, age 47 years (**d3**) (From Yurov et al. 2009b. Reproduced with permission of Oxford University Press in the format reuse in a book/textbook via Copyright Clearance Center)

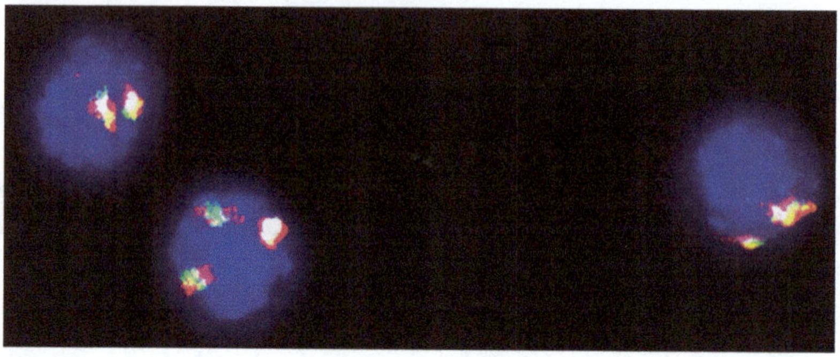

Fig. 4.7 Two nuclei with disomy 21 and a nucleus with true trisomy 21 revealed by ICS-MCB with chromosome 21-specific probe in the Alzheimer's disease (AD) brain (From Yurov et al. 2009a. Reproduced with permission of Academic Press in the format reuse in a book/textbook via Copyright Clearance Center)

cell death (Yang et al. 2001). Although single-color FISH allows analysis of DNA replication, some notes should be made, especially in relationship to postmitotic tissues. The best results of DNA replication activity in interphase nuclei are obtained by application of site-specific cosmid DNA probes for euchromatic chromosomal regions (Soloviev et al. 1995), whereas cosmid contig and centromeric DNA probes (used for studying AD brain) give contradictory results and have to be controlled by additional molecular cytogenetic techniques (Vorsanova et al. 2010a). Furthermore, more efficient molecular cytogenetic technologies have shown that tetraploid cells are really present in the AD brain (Mosch et al. 2007; Iourov et al. 2009b; Westra et al. 2009). However, Westra and coauthors have shown that these tetraploid nuclei are exclusively nonneuronal and are as prevalent as in the control (Westra et al. 2009; Chun et al. 2011; Iourov et al. 2011). An independent monitoring of aneuploidy/tetraploidy in the normal and AD brain by interphase mFISH has estimated true tetraploidy to affect 0.1–0.2 % of neural nuclei (Iourov et al. 2009b). These findings provide evidence against the relationship between tetraploidy and neurodegeneration. The paradoxes surrounding the AD cell-cycle theory arise from discrepancies between reproducible evidence for the presence of neurons exhibiting G_2 biomarkers and evidence against tetraploid genomic content in these neurons. To solve this paradox, the DNA replication stress hypothesis of AD was proposed (Yurov et al. 2011a). Accordingly, neurons entering into S-phase do not proceed further through the cell cycle and contain partially duplicated DNA content (Fig. 4.8). This finding suggests neuronal cell dysfunction and death occurs during the S-phase and originates from replication stress. In other words, unscheduled and unrealized DNA synthesis in vulnerable neurons, which epigenetically are unable to reorganize the nuclear genome for proper chromosome duplication, should lead to a DNA replication catastrophe or neuronal death resulting from lethal errors in replication. In this context, G_2-phase biomarkers are likely to be a sign of cell-cycle

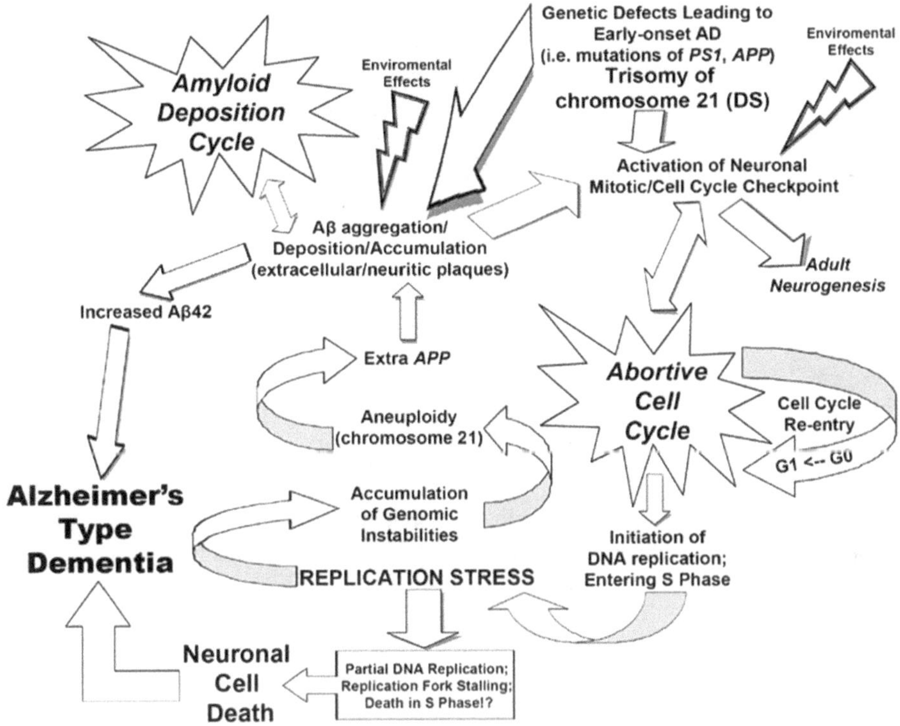

Fig. 4.8 Replication stress hypothesis of AD. Interplay between essential elements of the AD-type dementia pathogenetic cascade is proposed. The genetic influences (*PSEN* or *APP* mutations, trisomy 21, APOE4 genotype), metabolic changes, and environmental factors affecting neuronal homeostasis in the aging brain lead to activation of neuronal proliferation. Mitogens, which do exist in the human brain (neuronal cells), induce additional stimuli of extensive adult neurogenesis in the hippocampus. In the AD brain, such events would lead to increased hippocampal neurogenesis. A side effect could be that these mitogenic stimuli activate cell-cycle reentry in postmitotic neurons. The latter is a pathological activation of the neuronal cell cycle, including reentry into G_1- and S-phases and initiation of DNA replication. Neurons showing protein markers of G_2/M-phase probably contain a chromosome set of 23 duplicated chromosome pairs with unseparated chromatids (DNA content, 4C; chromosome complement, 2N) and become tetraploid in a sense of DNA content (4C). According to the commonly accepted theory of neuronal cell-cycle reentry and death, some neuronal populations complete the DNA synthesis but are arrested during the G_2/M transition. Therefore, neuronal death occurs in the G_2-phase. Alternatively, one can propose that a large proportion of activated postmitotic neurons in the AD brain are unable to pass the S-phase properly; this would lead to accumulation of genomic and chromosomal instabilities throughout ontogeny (DNA breaks, aneuploidy). In addition, replication-induced DNA damages would lead to fork stalling, incomplete or inefficient DNA replication, together designated as replication stress. Replication stress may be considered the leading cause of neuronal cell death caused by processing into S-phase or accumulation of genetic instabilities, which together constitute an important element of the AD pathogenetic cascade (From Yurov et al. 2011a. An open-access article distributed under the terms of the Creative Commons Attribution License)

"imitation" or other intracellular phenomena accompanied by production of G_2-specific proteins playing a role in processes of DNA repair, DNA damage response, and initiation of programmed cell death, but indirectly related to replicative cell-cycle events. Replication stress is a probable trigger of genome instability in the AD brain, which links abnormal cell-cycle events, chromosomal aneuploidy, and amyloid overproduction and deposition. Testing of the "replication stress—replicative death" hypothesis would help to expand our views on how neural cell-cycle dysregulation and somatic genome instability are involved in AD pathogenesis. Furthermore, such investigation can provide a clue to the role that genome instability plays in the normal and diseased brain in addition to the way genome stability is maintained in neuronal cells through ontogeny.

Origins of Aneuploidy in the Human Brain

The early stages of human embryonic development are prone to errors that produce aneuploidy or other types of somatic genome variations manifesting at the chromosomal level (Vorsanova et al. 2005, 2010a; Iourov et al. 2006c, 2010; Hassold et al. 2007; Dierssen et al. 2009; Robberecht et al. 2010; Yurov et al. 2010). Somatic genome instability including mosaic aneuploidy is extremely frequent among human embryos (Vanneste et al. 2009). Interphase FISH indicates that low-grade mosaic aneuploidy affecting more than 5–20 % of cells is frequently associated with spontaneous abortions being observed in 25 % of cases (Vorsanova et al. 2005). Therefore, low-level mosaicism is likely not to lead to prenatal death (Iourov et al. 2008a), which is supported by observations of somatic genome variations at chromosomal level in fetal tissues at 9–12 weeks of gestation (Yurov et al. 2007a). Together, these results suggest that global mitotic instability associated with aneuploidization in human fetal tissues is the main source of aneuploidy confined to the brain. Furthermore, embryonic neural cells have an extremely large number of mitotic divisions during early brain development (~250,000 cells per minute) (Muotri and Gage 2006), which can also be a reason for abundant brain aneuploidization because of mitotic machinery exhaustion in a dramatically accelerated cascade of cell divisions. Nonetheless, the intrinsic causes of aneuploidy in humans remain largely unknown (Iourov et al. 2006c, d, 2008a; Hassold et al. 2007; Li et al. 2009; Weaver and Cleveland 2009).

Aneuploidy increase in the diseased brain is likely to originate from natural cellular selection. This idea is further supported by observations that each disease exhibits chromosome-specific aneuploidy (chromosome-specific instability), for example, schizophrenia (chromosomes 1, 18, and X), Alzheimer's disease (chromosome 21), and ataxia-telangiectasia (chromosome 14) (Yurov et al. 2001, 2008; Iourov et al. 2009a, b). However, some of these are also associated with increased sporadic aneuploidy. Therefore, the selection is likely to be driven by different effects of alterations to cell clearance or "antianeuploidization" machinery (Iourov et al. 2008a). The extent of clearance failure determines the patterns of CIN or types of mosaic aneuploidy in the postnatal brain. A proportion of AD cases and ataxia-telangiectasia are known to be associated with mutations in specific genes. Thus, presenilin 1, which is mutated in early-onset familiar Alzheimer's disease, has been shown to cause chromosome missegregation and aneuploidy (Boeras et al. 2008). Amyloid precursor protein, an important element of the AD pathogenic cascade mutated in familiar AD, was also found to be involved in chromosome missegregation (Granic et al. 2010; Borysov et al. 2011). Finally, the mutated ataxia-telangiectasia gene (*ATM*), a component of genome integrity maintenance machinery involved in mitotic and apoptotic regulation, produces aneuploidy and chromosome-specific instability in the affected brain (Iourov et al. 2009a). Therefore, gene mutations can also contribute to formation of brain-specific aneuploidy.

Aneuploidy in the Aging Human Brain

Aneuploidy has been consistently shown to be associated with aging (Ly et al. 2000; Yurov et al. 2009b; Faggioli et al. 2011). However, the role of aneuploidy in the aging of the brain is largely unknown. An increasing rate of mitotic errors in late ontogeny can be a mechanism for chromosome gains and losses in aging tissues: this corresponds to data on aneuploidy in human tissues composed of mitotic cells but is not applicable to postmitotic neural cells. In this context, the human brain is probably the most remarkable example of a tissue populated by almost exclusively postmitotic cells that are not expected to undergo mitotic division.

Although somatic aneuploidy is associated with aging, the normal human brain is unlikely to feature a dramatic increase of aneuploidy rates during ontogeny (Iourov et al. 2008a). However, a reevaluation of aneuploidy in the postnatal human brain has shown aneuploidy rates tend to increase in this instance. The paradox has been theoretically solved proposing two scenarios: (1) natural cellular selection does not affect smaller populations of aneuploid cells, whereas the amount of euploid cells dramatically decreases throughout ontogeny; and (2) human adult neurogenesis and gliogenesis are prone to mitotic errors (Yurov et al. 2009b).

Mosaic neural aneuploidy is a remarkable biomarker of GIN and CIN. Looking through the data concerning aneuploidy in the developing and adult human central nervous system, the GIN 'n' CIN hypothesis of brain aging has been proposed, suggesting that neural aneuploidy produced during early brain development plays a crucial role of aging genetic determinant in the healthy and diseased brain (Yurov et al. 2009a). Key points of brain aging mediated by GIN/CIN are given in Fig. 4.9.

Interestingly, neurodegenerative diseases associated with abnormal/accelerated aging exhibit high rates of aneuploidy in the affected brain (Mosch et al. 2007; Arendt et al. 2009; Iourov et al. 2009a, b). To evaluate possible changes in the DNA content of brain cells during aging, Fischer et al. (2012) quantified the frequency of neurons with a more than diploid DNA content in the cerebral cortex of the normal human brain between the fourth and ninth decades of life. Their protocol included slide-based cytometry optimized for DNA quantification of single identified neurons, allowing DNA content analysis in about 500,000 neurons for each sample.

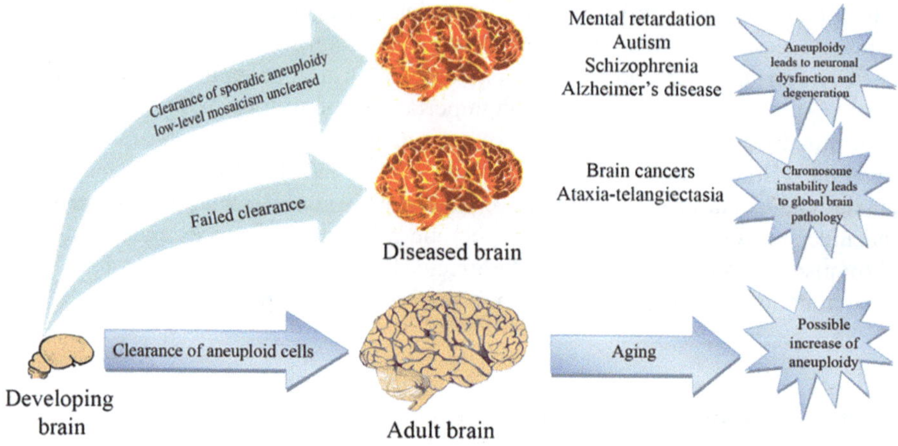

Fig. 4.9 Schematic representation of the hypothesis on the role of aneuploidy in normal central nervous system (CNS) development and aging as well as in pathogenesis of brain diseases. During normal prenatal brain development, developmental chromosome instability (CIN) is cleared, leading to threefold decrease of aneuploidy rates. Brain aging is likely to be associated with a slight increase of aneuploidy. Total failure of clearance of developmental CIN would lead to the persistence as observed in CIN syndromes with brain dysfunction (ataxia-telangiectasia) and brain cancers. Clearance may not affect low-level chromosomal mosaicism confined to the developing brain, which is extremely frequent among human fetuses. In such cases, the postnatal brain exhibits low-level chromosome-specific mosaic aneuploidy. The latter is shown to be associated with diseases of neuronal dysfunction and degeneration (mental retardation, autism, schizophrenia, Alzheimer's disease)

On average, 11.5 % of cortical neurons showed DNA content above the diploid level. The frequency of neurons with alterations to genomic content was highest in early adulthood/adolescence and declined with age. These results indicate that the genomic variation associated with DNA content exceeding the diploid level might compromise the viability of these neurons in the aging brain and might thus contribute to susceptibilities for age-related brain diseases. Alternatively, a potential selection bias of "healthy aging brains" needs to be considered, assuming that DNA content variation above a certain threshold associates with AD.

In contrast to DNA content variations in the aging human brain, the study of a mouse model provided alternative results. Faggioli et al. (2012) used the interphase FISH approach to compare aneuploidy levels in the aging murine brain. They showed that aneuploidy accumulates with age in a chromosome-specific manner (up to 9.8 % of nonneuronal brain nuclei in 28-month-old animals for chromosome 18). Although both neuronal and glial cells are affected equally at an early age, the age-related increase was limited to the nonneuronal nuclei. Extrapolating the data on average frequencies of aneuploidy involving 8 chromosomes to the entire murine genome (20 chromosomes), would indicate approximately 50 % cells of the aged murine brain to be aneuploid. Authors speculate that such high levels of genome instability affecting nonneuronal

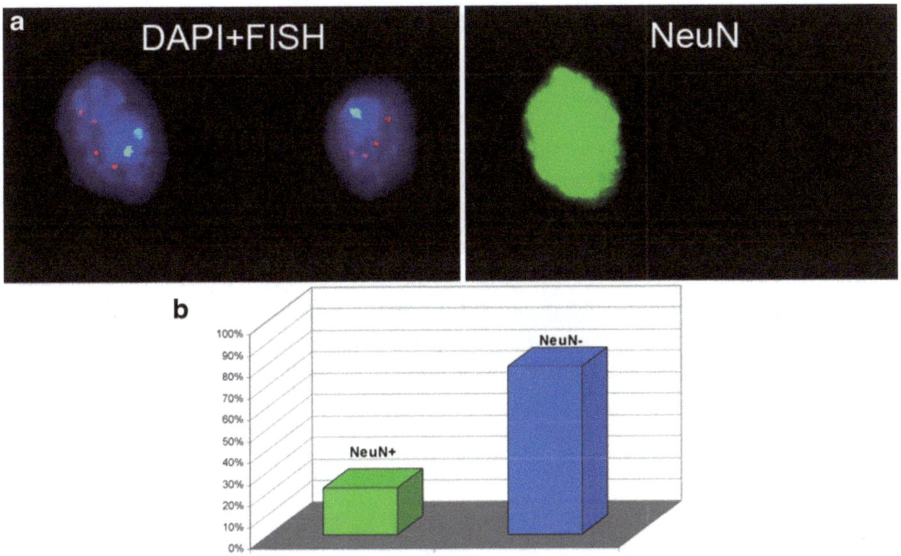

Fig. 4.10 Multicolor immuno-FISH (NeuN immunophenotyping+MFISH) of AT cerebellum cells. (**a** *left*): Simultaneous tricolor FISH with chromosome enumeration probes for chromosomes 1 (*blue signals*), 18 (*magenta signals*), and X (*red signals*) and DAPI staining demonstrate that one nucleus (*right*) is aneuploid (chromosome 1 loss). (**a** *right*): NeuN immunophenotyping of same nuclei demonstrates one NeuN-positive neuronal nucleus (*green color*, *left*) with two chromosomes 1 and one NeuN-negative aneuploid neuronal nucleus with monosomy 1. (**b**) Frequency of NeuN-positive and NeuN-negative neuronal-like nuclei with chromosomal imbalances in the cerebellum of an AT patient (From Yurov et al. 2009b. Reproduced with permission of Oxford University Press in the format reuse in a book/textbook via Copyright Clearance Center)

(glial) cells could be a cause of age-related neurodegeneration (Faggioli et al. 2012). This speculation is likely to correlate with analyses of aneuploidy in the AT brain (early-onset progressive neurodegenerative disease characterized by premature aging) (Iourov et al. 2009a). In this premature aging disease, increased aneuploidy and chromosome breaks in the brain were predominantly found in nonneuronal cells (up to 80 %) of the cerebellum (Fig. 4.10). Therefore, available data generally confirm the significance of somatic genome and CIN in the brain during late ontogeny or aging.

Interphase Chromosomes and Genome Organization in the Human Brain

The availability of technical solutions for studying interphase chromosomes in the human brain allows analyzing the nuclear genome organization as well (Vorsanova et al. 2010a; Iourov et al. 2006b, 2010, 2012). Although some previous efforts have provided for intriguing data on specific patterns of chromosome behavior (chromosomal associations, somatic pairing of homologous chromosome regions) and its probable contribution to brain diseases (for review, see Leitch 2000; Iourov et al. 2006c), this area of molecular neurocytogenetics remains almost unstudied. Apart from a few reports on associations of heterochromatic and much more rarely euchromatic regions (Arnoldus et al. 1989, 1991; Leitch 2000; Iourov et al. 2005, 2006a, 2010), chromosome dynamics and chromatin organization at the chromosomal level in interphase nuclei of human neuronal cells are almost completely unknown. Therefore, it seems that molecular neurocytogenetic analyses of functional interphase chromosome organization at the chromosomal and subchromosomal level are strongly required for filling the gaps in our knowledge of genome behavior in the human central nervous system.

Conclusion

The present review is aimed at describing the latest advances in molecular neurocytogenetics with special attention to chromosome (genome) variations in postmitotic cells of the human brain. Aneuploidy is considered as a highly pathogenic type of GIN. Mosaic aneuploidy in the brain is the result of mitotic cell-cycle errors during developmental and adult neurogenesis and, probably, gliogenesis. Paradoxically, addressing neurocytogenetic data, one can conclude that low-level constitutional aneuploidy is an integral component of normal human central nervous system development and could mediate neuronal diversity. Nevertheless, the pathogenetic cascade producing neural genome instability seems to increase neural aneuploidy rates in brain diseases. The role of aneuploidy, tetraploidy, and ectopic DNA replication events in the brain is the basis for numerous hypotheses. Taking into account that some neurodegenerative diseases exhibiting acquired brain-specific aneuploidy are those associated with pathological or accelerated aging, speculations about relationship between "nonmalignant aneuploidization," neurodegeneration, and brain aging are pertinent.

The main outcome of previous molecular neurocytogenetic studies is that mosaic aneuploidy does affect the developing and adult human brain. In the developing human brain, aneuploidy is likely to regulate cell numbers and is probably a kind of "checkpoint" for programmed cell death. In the adult human brain, aneuploid cells are likely to represent a signature of developmental CIN. One still cannot exclude that aneuploidy also plays a role in human neuronal diversity. The lack of clearance of aneuploid cells is likely to be a mechanism for human brain diseases associated with CIN and low-level mosaic aneuploidy in the brain. However, the origins of aneuploidy and its effects on cellular physiology remain to be established. Furthermore, there are psychiatric and neurological disorders that require direct studies of genome variability and instability in the diseased brain. We propose that current experimental evidence and attractive (but untested) hypotheses concerning genome variation in the brain can be used for proposing a theory of neural genome ontogenetic instability in health and disease. This theory would explain the role of somatic genome variation in the etiology and pathogenesis of brain diseases and, probably, in both normal and pathological brain aging. Finally, it is pointed out that molecular neurocytogenetics and cytogenomics are integral parts of current biomedicine and possess the potential to yield new discoveries in human genetics, genomics, neuroscience, and cell biology.

Acknowledgments The review is dedicated to the memory of Dr. Ilya V. Soloviev. We gratefully acknowledge the NICHD Brain and Tissue Bank for Developmental Disorders at the University of Maryland, Baltimore, MD, USA, for providing samples of brain tissues. The authors are supported by DLR/BMBF (RUS 2011–2013) and RFBR grant 12-04-00215- (Russian Federation, 2012–2014). Dr. IY Iourov is supported by the Grant of the President of the Russian Federation MD-4401.2013.7.

References

Arendt T (2012) Cell cycle activation and aneuploid neurons in Alzheimer's disease. Mol Neurobiol 46(1):125–135

Arendt T, Mosch B, Morawski M (2009) Neuronal aneuploidy in health and disease: a cytomic approach to understand the molecular individuality of neurons. Int J Mol Sci 10:1609–1627

Arendt T, Brückner MK, Mosch B, Lösche A (2010) Selective cell death of hyperploid neurons in Alzheimer's disease. Am J Pathol 177:15–20

Arnoldus EP, Peters AC, Bots GT, Raap AK, van der Ploeg M (1989) Somatic pairing of chromosome 1 centromeres in interphase nuclei of human cerebellum. Hum Genet 83(3):231–234

Arnoldus EP, Noordermeer IA, Peters AC, Raap AK, Van der Ploeg M (1991) Interphase cytogenetics reveals somatic pairing of chromosome 17 centromeres in normal human brain tissue, but no trisomy 7 or sex-chromosome loss. Cytogenet Cell Genet 56(3–4):214–216

Arnoldus EP, Wolters LB, Voormolen JH, van Duinen SG, Raap AK, van der Ploeg M, Peters AC (1992) Interphase cytogenetics: a new tool for the study of genetic changes in brain tumors. J Neurosurg 76(6):997–1003

Astolfi PA, Salamini F, Sgaramella V (2010) Are we genomic mosaics? Variations of the genome of somatic cells can contribute to diversify our phenotypes. Curr Genomics 11:379–386

Betancur C (2011) Etiological heterogeneity in autism spectrum disorders: more than 100 genetic and genomic disorders and still counting. Brain Res 1380:42–77

Boeras DI, Granic A, Padmanabhan J, Crespo NC, Rojiani AM, Potter H (2008) Alzheimer's presenilin 1 causes chromosome missegregation and aneuploidy. Neurobiol Aging 29:319–328

Borysov SI, Granic A, Padmanabhan J, Walczak CE, Potter H (2011) Alzheimer Aβ disrupts the mitotic spindle and directly inhibits mitotic microtubule motors. Cell Cycle 10:1397–1410

Castermans D, Willquet V, Steyert J, Van de Ven W, Fryns JP, Devriendt K (2004) Chromosomal anomalies in individuals with autism: a strategy towards the identification of genes involved in autism. Autism 8:141–161

Chen J, Cohen ML, Lerner AJ, Yang Y, Herrup K (2010) DNA damage and cell cycle events implicate cerebellar dentate nucleus neurons as targets of Alzheimer's disease. Mol Neurodegener 5:60

Chun J, Westra JW, Bushman D (2011) Reply to Iourov et al. Neurodegen Dis 8:38–40

Dastidar SG, Bardai FH, Ma C, Price V, Rawat V, Verma P, Narayanan V, D'Mello SR (2012) Isoform-specific toxicity of Mecp2 in postmitotic neurons: suppression of neurotoxicity by FoxG1. J Neurosci 32(8):2846–2855

de Moraes LS, Khayat AS, de Lima PD, Lima EM, Pinto GR, Leal MF, de Arruda Cardoso Smith M, Burbano RR (2010) Chromosome X aneuploidy in Brazilian schizophrenic patients. In Vivo 24:281–286

DeLisi LE, Friedrich U, Wahlstrom J, Boccio-Smith A, Eklund K, Crow TJ (1994) Schizophrenia and sex chromosome anomalies. Schizophr Bull 20(3):495–505

DeLisi LE, Maurizio AM, Svetina C, Ardekani B, Szulc K, Nierenberg J et al (2005) Klinefelter's syndrome (XXY) as a genetic model for psychotic disorders. Am J Med Genet B Neuropsychiatr Genet 135B(1):15–23

Dierssen M, Herault Y, Estivill X (2009) Aneuploidy: from a physiological mechanism of variance to Down syndrome. Physiol Rev 89:887–920

Duesberg P, Li R, Fabarius A, Hehlmann R (2005) The chromosomal basis of cancer. Cell Oncol 27(5-6):293–318

Faggioli F, Vijg J, Montagna C (2011) Chromosomal aneuploidy in the aging brain. Mech Ageing Dev 132(8–9):429–436

Faggioli F, Wang T, Vijg J, Montagna C (2012) Chromosome-specific accumulation of aneuploidy in the aging mouse brain. Hum Mol Genet 21(24):5246–5253. doi:10.1093/hmg/dds375

Fischer HG, Morawski M, Brückner MK, Mittag A, Tarnok A, Arendt T (2012) Changes in neuronal DNA content variation in the human brain during aging. Aging Cell 11(4):628–633

Geller LN, Potter H (1999) Chromosome missegregation and trisomy 21 mosaicism in Alzheimer's disease. Neurobiol Dis 6:167–179

Granic A, Padmanabhan J, Norden M, Potter H (2010) Alzheimer Abeta peptide induces chromosome mis-segregation and aneuploidy, including trisomy 21: requirement for au and APP. Mol Biol Cell 21(4):511–520

Hassold T, Hall H, Hunt P (2007) The origin of human aneuploidy: where we have been, where we are going. Hum Mol Genet 16:R203–R208

Herrup K, Yang Y (2007) Cell cycle regulation in the postmitotic neuron: oxymoron or new biology? Nat Rev Neurosci 8:368–378

Heston LL, Mastri AR (1977) The genetics of Alzheimer's disease: associations with hematologic malignancy and Down's syndrome. Arch Gen Psychiatry 34(8):976–981

Hultén MA, Patel SD, Tankimanova M, Westgren M, Papadogiannakis N, Jonsson AM, Iwarsson E (2008) On the origin of trisomy 21 Down syndrome. Mol Cytogenet 1:21

Hultén MA, Jonasson J, Nordgren A, Iwarsson E (2010) Germinal and somatic trisomy 21 mosaicism: how common is it, what are the implications for individual carriers and how does it come about? Curr Genomics 11:409–419

Hultén MA, Jonasson J, Iwarsson E, Uppal P, Vorsanova SG, Yurov YB, Iourov IY (2013) Trisomy 21 mosaicism: we may all have a touch of Down syndrome. Cytogenet Genome Res. DOI:10.1159/000346028

Iourov IY, Soloviev IV, Vorsanova SG, Monakhov VV, Yurov YB (2005) An approach for quantitative assessment of fluorescence in situ hybridization (FISH) signals for applied human molecular cytogenetics. J Histochem Cytochem 53:401–408

Iourov IY, Liehr T, Vorsanova SG, Kolotii AD, Yurov YB (2006a) Visualization of interphase chromosomes in postmitotic cells of the human brain by multicolour banding (MCB). Chromosome Res 14:223–229

Iourov IY, Vorsanova SG, Pellestor F, Yurov YB (2006b) Brain tissue preparations for chromosomal PRINS labeling. Methods Mol Biol 334:123–132

Iourov IY, Vorsanova SG, Yurov YB (2006c) Chromosomal variation in mammalian neuronal cells: known facts and attractive hypotheses. Int Rev Cytol 249:143–191

Iourov IY, Vorsanova SG, Yurov YB (2006d) Intercellular genomic (chromosomal) variations resulting in somatic mosaicism: mechanisms and consequences. Curr Genomics 7:435–446

Iourov IY, Liehr T, Vorsanova SG, Yurov YB (2007a) Interphase chromosome-specific multicolor banding (ICS-MCB): a new tool for analysis of interphase chromosomes in their integrity. Biomol Eng 24:415–417

Iourov IY, Vorsanova SG, Yurov YB (2007b) Ataxia telangiectasia paradox can be explained by chromosome instability at the subtissue level. Med Hypotheses 68:716

Iourov IY, Vorsanova SG, Yurov YB (2008a) Chromosomal mosaicism goes global. Mol Cytogenet 1:26

Iourov IY, Vorsanova SG, Yurov YB (2008b) Molecular cytogenetics and cytogenomics of brain diseases. Curr Genomics 9:452–465

Iourov IY, Yurov YB, Vorsanova SG (2008c) Mosaic X chromosome aneuploidy can help to explain the male-to-female ratio in autism. Med Hypotheses 70:456

Iourov IY, Vorsanova SG, Liehr T, Kolotii AD, Yurov YB (2009a) Increased chromosome instability dramatically disrupts neural genome integrity and mediates cerebellar degeneration in the ataxia-telangiectasia brain. Hum Mol Genet 18:2656–2669

Iourov IY, Vorsanova SG, Liehr T, Yurov YB (2009b) Aneuploidy in the normal, Alzheimer's disease and ataxia-telangiectasia brain: differential expression and pathological meaning. Neurobiol Dis 34:212–220

Iourov IY, Vorsanova SG, Yurov YB (2009c) Developmental neural chromosome instability as a possible cause of childhood brain cancers. Med Hypotheses 72:615–616

Iourov IY, Vorsanova SG, Yurov YB (2010) Somatic genome variations in health and disease. Curr Genomics 11:387–396

Iourov IY, Vorsanova SG, Yurov YB (2011) Genomic landscape of the Alzheimer's disease brain: chromosome instability—aneuploidy, but not tetraploidy—mediates neurodegeneration. Neurodegener Dis 8:35–37

Iourov IY, Vorsanova SG, Yurov YB (2012) Single cell genomics of the brain: focus on neuronal diversity and neuropsychiatric diseases. Curr Genomics 13(6):477–488

Kingsbury MA, Friedman B, McConnell MJ, Rehen SK, Yang AH, Kaushal D, Chun J (2005) Aneuploid neurons are functionally active and integrated into brain circuitry. Proc Natl Acad Sci U S A 102:6143–6147

Kingsbury MA, Yung YC, Peterson SE, Westra JW, Chun J (2006) Aneuploidy in the normal and diseased brain. Cell Mol Life Sci 63(22):2626–2641

Leitch AR (2000) Higher levels of organization in the interphase nucleus of cycling and differentiated cells. Microbiol Mol Biol Rev 64(1):138–152

Li L, McCormack AA, Nicholson JM, Fabarius A, Hehlmann R, Sachs RK, Duesberg PH (2009) Cancer-causing karyotypes: chromosomal equilibria between destabilizing aneuploidy and stabilizing selection for oncogenic function. Cancer Genet Cytogenet 188:1–25

Ly DH, Lockhar DJ, Lerne RA, Schultz PG (2000) Mitotic misregulation and human aging. Science 287:2486–2492

McConnell MJ, Kaushal D, Yang AH, Kingsbury MA, Rehen SK, Treuner K et al (2004) Failed clearance of aneuploid embryonic neural progenitor cells leads to excess aneuploidy in the Atm-deficient but not the Trp53-deficient adult cerebral cortex. J Neurosci 24:8090–8096

McKinnon PJ (2004) ATM and ataxia telangiectasia. EMBO Rep 5:772–776

Migliore L, Coppede F, Fenech M, Thomas P (2011) Association of micronucleus frequency with neurodegenerative diseases. Mutagenesis 26:85–92

Mosch B, Morawski M, Mittag A, Lenz D, Tarnok A, Arendt T (2007) Aneuploidy and DNA replication in the normal human brain and Alzheimer's disease. J Neurosci 27:6859–6867

Muotri AR, Gage FH (2006) Generation of neuronal variability and complexity. Nature (Lond) 441:1087–1093

Potter H (2008) Down's syndrome and Alzheimer's disease: two sides of the same coin. Future Neurol 3:29–37

Potter H, Granic A, Iourov IY, Migliore L, Vorsanova SG, Yurov YB (2011) Alzheimer's insight. The New Scientist 211(2824):32

Rehen SK, McConnell MJ, Kaushal D, Kingsbury MA, Yang AH, Chun J (2001) Chromosomal variation in neurons of the developing and adult mammalian nervous system. Proc Natl Acad Sci U S A 98:13361–13366

Rehen SK, Yung YC, McCreight MP, Kaushal D, Yang AH, Almeida BSV et al (2005) Constitutional aneuploidy in the normal human brain. J Neurosci 25(9):2176–2180

Robberecht C, Vanneste E, Pexsters A, D'Hooghe T, Voet T, Vermeesch JR (2010) Somatic genomic variations in early human prenatal development. Curr Genomics 11(6):397–401

Smith CL, Bolton A, Nguyen G (2010) Genomic and epigenomic instability, fragile sites, schizophrenia and autism. Curr Genomics 11(6):447–469

Soloviev IV, Yurov YB, Vorsanova SG, Fayet F, Roizes G, Malet P (1995) Prenatal diagnosis of trisomy 21 using interphase fluorescence in situ hybridization of post-replicated cells with site-specific cosmid and cosmid contig probes. Prenat Diagn 15:237–248

Spremo-Potrapevic B, Zivkovic L, Plecas-Solarovic B, Bajic VP (2011) Chromosome instability in Alzheimer's disease. Arch Biol Sci 63:603–608

Taupin P (2011) Neurogenesis, NSCs, pathogenesis and therapies for Alzheimer's disease. Front Biosci 3:178–190

Thomas P, Fenech M (2008) Chromosome 17 and 21 aneuploidy in buccal cells is increased with ageing and in Alzheimer's disease. Mutagenesis 23:57–65

Vanneste E, Voet T, Le Caignec C, Ampe M, Konings P, Melotte C et al (2009) Chromosome instability is common in human cleavage-stage embryos. Nat Med 15:577–583

Vorsanova SG, Demidova IA, Ulas VY, Soloviev IV, Kazantzeva LZ, Yurov YB (1996) Cytogenetic and molecular-cytogenetic investigation of Rett syndrome: analysis of 31 cases. Neuroreport 8(1):187–189

Vorsanova SG, Yurov YB, Ulas VY, Demidova IA, Sharonin VO, Kolotii AD et al (2001) Cytogenetic and molecular-cytogenetic studies of Rett syndrome (RTT): a retrospective analysis of a Russian cohort of RTT patients (the investigation of 57 girls and three boys). Brain Dev 23:S196–S201

Vorsanova SG, Kolotii AD, Iourov IY, Monakhov VV, Kirillova EA, Soloviev IV, Yurov YB (2005) Evidence for high frequency of chromosomal mosaicism in spontaneous abortions revealed by interphase FISH analysis. J Histochem Cytochem 53:375–380

Vorsanova SG, Yurov IY, Demidova IA, Voinova-Ulas VY, Kravets VS, Solov'ev IV et al (2007) Variability in the heterochromatin regions of the chromosomes and chromosomal anomalies in children with autism: identification of genetic markers of autistic spectrum disorders. Neurosci Behav Physiol 37:553–558

Vorsanova SG, Iourov IY, Kolotii AD, Beresheva AK, Demidova IA, Kurinnaia OS et al (2010a) Chromosomal mosaicism in spontaneous abortions: analysis of 650 cases. Russ J Genet 46:1197–1200

Vorsanova SG, Voinova VY, Yurov IY, Kurinnaya OS, Demidova IA, Yurov YB (2010b) Cytogenetic, molecular-cytogenetic, and clinical-genealogical studies of the mothers of children with autism: a search for familial genetic markers for autistic disorders. Neurosci Behav Physiol 40(7):745–756

Vorsanova SG, Yurov YB, Iourov IY (2010c) Human interphase chromosomes: a review of available molecular cytogenetic technologies. Mol Cytogenet 3:1

Weaver BA, Cleveland DW (2009) The role of aneuploidy in promoting and suppressing tumors. J Cell Biol 185:935–937

Weier HU, Munne S, Ferlatte C, Baumgartner A, Iourov IY, Racowsky C et al (2010) Aneuploidy—a necessary evil in human life. In: New trends in microscopy & immunohistochemistry (Histochemistry 2010), Proceedings of the histochemical 61st annual meeting. The Marine Biological Laboratory, Woods Hole, MA, pp 42–43

Westra JW, Peterson SE, Yung YC, Mutoh T, Barral S, Chun J (2008) Aneuploid mosaicism in the developing and adult cerebellar cortex. J Comp Neurol 507:1944–1951

Westra JW, Barral S, Chun J (2009) A reevaluation of tetraploidy in the Alzheimer's disease brain. Neurodegener Dis 6:221–229

Westra JW, Rivera RR, Bushman DM, Yung YC, Peterson SE, Barral S, Chun J (2010) Neuronal DNA content variation (DCV) with regional and individual differences in the human brain. J Comp Neurol 518:3981–4000

Xu J, Zwaigenbaum L, Szatmari P, Scherer SW (2004) Molecular cytogenetics of autism. Curr Genomics 4:347–368

Yang Y, Herrup K (2007) Cell division in the CNS: protective response or lethal event in post-mitotic neurons? Biochim Biophys Acta 1772:457–466

Yang Y, Geldmacher DS, Herrup K (2001) DNA replication precedes neuronal cell death in Alzheimer's disease. J Neurosci 21:2661–2668

Yurov YB, Vostrikov VM, Vorsanova SG, Monakhov VV, Iourov IY (2001) Multicolor fluorescent in situ hybridization on post-mortem brain in schizophrenia as an approach for identification of low-level chromosomal aneuploidy in neuropsychiatric diseases. Brain Dev 23:S186–S190

Yurov YB, Iourov IY, Monakhov VV, Soloviev IV, Vostrikov VM, Vorsanova SG (2005) The variation of aneuploidy frequency in the developing and adult human brain revealed by an interphase FISH study. J Histochem Cytochem 53:385–390

Yurov YB, Iourov IY, Vorsanova SG, Liehr T, Kolotii AD, Kutsev SI et al (2007a) Aneuploidy and confined chromosomal mosaicism in the developing human brain. PLoS One 2:e558

Yurov YB, Vorsanova SG, Iourov IY, Demidova IA, Beresheva AK, Kravetz VS et al (2007b) Unexplained autism is frequently associated with low-level mosaic aneuploidy. J Med Genet 44:521–525

Yurov YB, Iourov IY, Vorsanova SG, Demidova IA, Kravetz VS, Beresheva AK et al (2008) The schizophrenia brain exhibits low-level aneuploidy involving chromosome 1. Schizophr Res 98:139–147

Yurov YB, Iourov IY, Vorsanova SG (2009a) Neurodegeneration mediated by chromosome instability suggests changes in strategy for therapy development in ataxia-telangiectasia. Med Hypotheses 73:1075–1076

Yurov YB, Vorsanova SG, Iourov IY (2009b) GIN 'n' CIN hypothesis of brain aging: deciphering the role of somatic genetic instabilities and neural aneuploidy during ontogeny. Mol Cytogenet 2:23

Yurov YB, Vorsanova SG, Iourov IY (2010) Ontogenetic variation of the human genome. Curr Genomics 11:420–425

Yurov YB, Vorsanova SG, Iourov IY (2011a) The DNA replication stress hypothesis of Alzheimer's disease. ScientificWorldJournal 11:2602–2612

Yurov YB, Vorsanova SG, Kolotii AD, Liehr T, Iourov IY (2011b) Aneuploidy in the autistic brain: the first molecular cytogenetic study. Balkan J Med Genet 14(suppl 1):73

Yurov YB, Vorsanova SG, Kolotii AD, Liehr T, Iourov IY (2012) Brain-specific X chromosome aneuploidy is likely to contribute to the pathogenesis of autism and can explain the unsolved paradox of male susceptibility. Eur J Hum Genet 20(suppl 1):109

Zekanowski C, Wojda U (2009) Aneuploidy, chromosomal missegregation, and cell cycle reentry in Alzheimer's disease. Acta Neurobiol Exp 6:232–253

Chapter 5
Differentiating Chromosome Fragmentation and Premature Chromosome Condensation

Joshua B. Stevens and Henry H.Q. Heng

Abstract The chromosome has long been viewed as a structure that ensures faithful segregation of the genetic materials to daughter cells. However, it is now apparent that the chromosome plays a central role in defining the genetic network through the genome context. One often-confused phenomenon bridging studies of interphase chromatin and mitotic chromosomes is chromosome pulverization, which has been inappropriately linked to premature chromosome condensation (PCC) and more recently confused with chromosome fragmentation (C-Frag), a major form of mitotic cell death. Recently there has been increased interest in genome alteration-mediated somatic cell evolution and its clinical implications, although a number of publications have continued to confuse these terminologies/concepts.

To alleviate confusion in this field we review both C-Frag and PCC. Discussion of C-Frag includes its morphological and mechanistic characterization, its relationship to genomic instability, and its utility. Discussion of PCC pertains to its mechanisms, definition, historical perspectives, and its application in basic research and clinical settings. C-Frag and PCC are then directly compared and contrasted to fully differentiate these two phenomena. Chromosome pulverization, chromosome shattering, and mitotic catastrophe are compared in relationship to both C-Frag and PCC. To avoid future confusion we suggest avoidance of the ambiguous term chromosome pulverization in

J.B. Stevens
Center for Molecular Medicine and Genetics, Wayne State University
School of Medicine, Detroit, MI, USA

H.H.Q. Heng (✉)
Center for Molecular Medicine and Genetics, Wayne State University
School of Medicine, Detroit, MI, USA

Department of Pathology, Wayne State University School of Medicine,
Detroit, MI, USA

Karmanos Cancer Institute, Detroit, MI, USA
e-mail: hheng@med.wayne.edu

Y.B. Yurov et al. (eds.), *Human Interphase Chromosomes: Biomedical Aspects*,
DOI 10.1007/978-1-4614-6558-4_5, © Springer Science+Business Media, LLC 2013

favor of the more specific term C-Frag or PCC. Finally, future implications and perspectives of both C-Frag and PCC are discussed.

Introduction

It is well established that the chromosome plays a key role in packaging the genes to ensure faithful division of genetic material during mitosis (Heng et al. 2004). However, the chromosome serves more than this obvious purpose (Heng et al. 2011a). Recently, it has been realized that the order of genes along the chromosome and within the genome represent a new type of genetic information called system inheritance. The genome provides this new information through the genome topology, which is an important component of the genome context (all sequences of a given species plus genomic topology), which in turn defines the genetic network (Heng et al. 2009, 2011a, b; Heng 2013). Thus, the most important functions of the chromosomes are (1) defining the genetic network for all types of somatic cells, and (2) ensuring the maintenance of system inheritance (especially through the germline) by preserving the karyotype (including the order of genes along the chromosome, as well as the chromosomal compositions within a cell). As the result, alterations in chromosomal number or structure lead to extensive changes in gene expression, modifying the networks in which those genes function (Stevens et al. 2013a, b). Maintenance of chromosome makeup (both structurally and numerically) of a cell is thus of great importance, and alteration of this makeup results in multiple diseases, especially cancer, by providing evolutionary potential. Therefore, there is an urgent need to increase understanding of the chromosome and the effects of its alteration (Heng et al. 2009, 2010a, 2013a; Heng 2007, 2009; Stevens et al. 2011a; Gorelick and Heng 2011).

Premature chromosome condensation (PCC) is a phenomenon whereby chromatin condensation is induced inappropriately during interphase. The discovery of PCC has led to increased knowledge of the basic chromosome structure, the state of the chromosomes during the different stages of interphase, and identification of factors involved in the cell cycle (Bezrookove et al. 2003; Johnson and Rao 1970). One key feature of PCC is the resultant chromosomal morphology (Potu et al. 1977). Induction of PCC during S-phase results in the condensation of partially replicated chromosomes, which appear as fragmented clumps of chromosomes. Interestingly, similar morphology, often described as chromosome pulverization or shattering, has been observed following exposure to a variety of agents, including viral infection, pesticides, caffeine, and ultraviolet light, and in blood diseases (Knuutila et al. 1981; Alam and Kasatiya 1976; Cremer et al. 1980; Norrby et al. 1966; Kato and Sandberg 1968). Despite the difference of pulverization from PCC in many of these cases, and warnings of restraint in calling PCC chromosome pulverization, the use of the term pulverization has remained (Stevens et al. 2007, 2010, 2011a, b; Sandberg 1978).

More recently, a new form of mitotic cell death called chromosome fragmentation (C-Frag) has been identified (Heng et al. 2004, 2013b; Stevens et al. 2007, 2011a, 2013b; Ye et al. 2007). During chromosome fragmentation, condensed mitotic chromosomes are progressively degraded, leading to cell death. C-Frag is induced to eliminate cells subjected to broad-ranging stress, including both external stress such as drug treatments or viral infections and internal stress including genome instability (Stevens et al. 2011a). Interestingly, incomplete chromosome fragmentation leads to fragmented pieces of chromosomes being rejoined to form highly complex chromosomal rearrangements known as genome chaos and a subtype of genome chaos, chromothripsis (Heng et al. 2006, 2009, 2011a, b; Kloosterman et al. 2011; Liu et al. 2013; Stephens et al. 2011). Thus, C-Frag can act as a double-edged sword, on one hand serving to retain genomic integrity by removing stressed and altered cells and on the other hand changing the genome and perpetuating somatic evolution.

Because of the similarities between C-Frag and PCC, these two phenomena are frequently confused. In particular, such confusion is evidenced in many recent studies seeking to understand the mechanism of genome chaos (Heng et al. 2010b, 2011a; Liu et al. 2013; Micronuclear chromosome pulverization may underlie chromothripsis 2012; Crasta et al. 2012). This review seeks to quell that confusion. To do so, C-Frag and PCC are discussed, specifically in regard to similarities and differences in morphology, mechanisms, and outcomes. The basic research and clinical utilities of both phenomena are also discussed. Similarities of the two to other confusing phenomena such as mitotic catastrophe, chromosome shattering, and chromosome pulverization are briefly addressed. Finally, future avenues of research and implications of these studies are described to provide a broad view and underscore the importance of both C-Frag and PCC.

Chromosome Fragmentation

C-Frag Is a Major Type of Mitotic Cell Death

C-Frag is a major form of mitotic cell death that occurs directly during mitosis, both in vivo and in vitro, and results in the progressive degradation of condensed, mitotic chromosomes (Figs. 5.1 and 5.2) (Stevens et al. 2007). C-Frag has been previously observed but was often referred to as PCC (Stevens et al. 2007, 2010). The realization that C-Frag resulted from the degradation of mitotic chromosomes resulted in the conclusion that C-Frag and PCC are distinct mechanisms (Stevens et al. 2007, 2010, 2011a). In addition to the distinctive morphology of C-Frag, degraded chromosomes are mitotic, as evidenced by phosphorylation of histone H3 at Ser10, whereas viability is lost during the degradation of chromosomes, as well as when they are exclusively induced from the window of mitosis (Stevens et al. 2007).

C-Frag differs from apoptosis in both morphology and mechanism. Morphologically, on cytogenetic slides apoptotic cells appear as clusters of small round regions of condensed DNA (Fig. 5.3) (Stevens et al. 2007). Mechanistically, C-Frag is not affected by overexpression of Bcl-2 or caspase inhibition, and fragmented chromosomes have been shown to not react to TUNEL staining, although double-strand breaks are detectable during C-Frag by γ-H2AX (Stevens et al. 2007). Apoptosis, on the other hand, is inhibited by caspase inhibition and Bcl-2 overexpression, and degraded DNA from apoptotic cells exhibit positive TUNEL staining. Interestingly although C-Frag

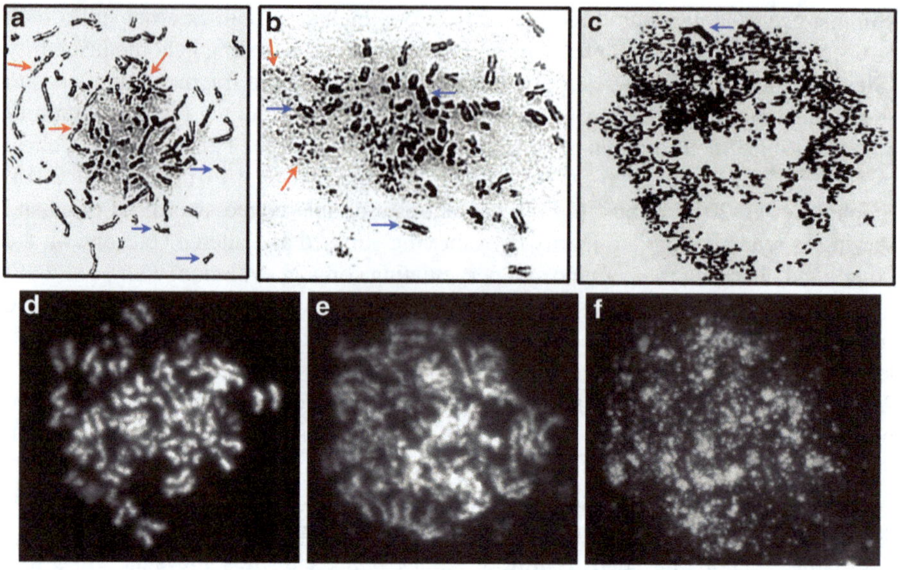

Fig. 5.1 Examples of various stages of chromosome fragmentation (C-Frag). Giemsa (**a**–**c**) and DAPI (**d**–**f**) images of early- (**a**, **d**), mid- (**b**, **e**), and late-stage (**c**, **f**) C-Frag. In early-stage C-Frag, a limited number of chromosomes begin to degrade, as evidenced by broken chromosomes (*red arrows*), while most remain unphased (*blue arrows*). As C-Frag continues, more chromosomes show signs of degradation and chromosome morphology begins to be lost. At late stages of C-Frag, nearly all chromosomes are degraded, although one or more chromosomes may still be intact (*blue arrow*) (**a**–**c** Adapted from Stevens et al. 2007)

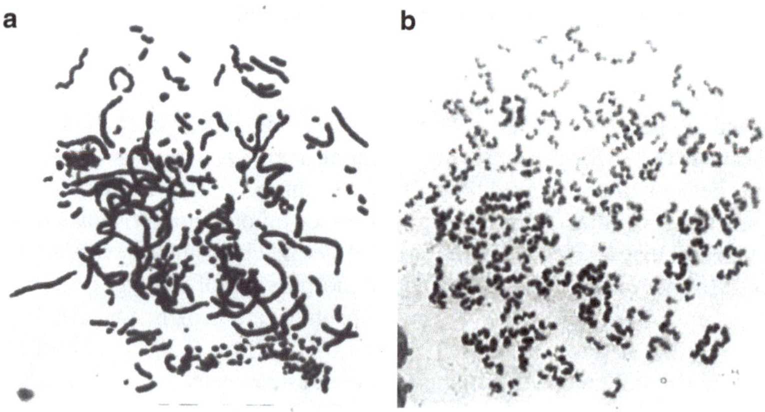

Fig. 5.2 C-Frag occurring at later stages of mitosis and earlier stages of mitosis. The morphology of C-Frag is also dependent on the stage of mitosis in which it occurs. C-Frag occurring early in mitosis (pre-metaphase) results in chromosomes that are not highly condensed and in which separation of sister chromatids is not detectable (**a**). Occurrence of C-Frag at or beyond metaphase results in further condensation of chromosomes, and separation of sister chromatids is evident (**b**). Occurrence of C-Frag in response to various drug treatments more commonly results in chromosome degradation at or after metaphase, especially if the treatment includes inhibitors of microtubule dynamics

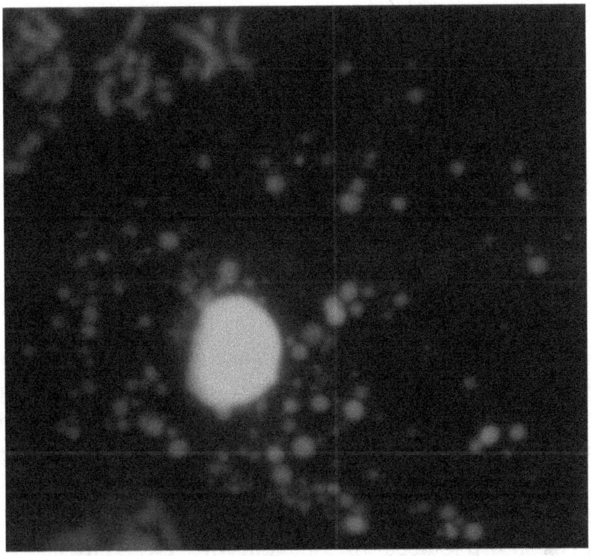

Fig. 5.3 Example of an apoptotic cell. Although apoptosis can result in condensation of fragmented DNA, this condensation differs in morphology from chromatin condensation. Apoptotic DNA fragments aggregate into small, condensed, circular clusters of DNA that differ drastically from condensed chromosomes

differs from apoptosis, poly ADP ribose polymerase (PARP) degradation does take place during fragmentation, indicating that C-Frag may be a programmed cell death where DNA repair by PARP is undesirable during the process (Stevens et al. 2011a).

C-Frag also differs in from mitotic catastrophe (MC), an apparently different form of mitotic cell death (Stevens et al. 2011a). Although the Committee on Cell Death Nomenclature has advised against the term mitotic catastrophe, studies reporting death by mitotic catastrophe are pervasive (Kroemer et al. 2005). Multiple reports have described MC as a cell death that occurs following an abnormal or abortive mitosis (Castedo et al. 2004a; Chan et al. 1999; Roninson et al. 2001). In some cases MC is linked to caspase activation (Castedo et al. 2004a). Morphologically, MC results in multilobulated nuclei and micronuclei that form during the death process (Castedo et al. 2004b). MC morphology is therefore easily distinguishable from C-Frag (Stevens et al. 2011a). Although C-Frag appears to be distinct from other forms of cell death, including apoptosis and MC, it is related to these other deaths in that it works in concert with them to ensure that abnormal and damaged cells are effectively eliminated (Stevens et al. 2011a).

Morphological Characterization of C-Frag

Broadly, C-Frag appears as three distinct morphological groupings: early C-Frag, where limited fragmentation has taken place on a limited number of chromosomes; late C-Frag, where nearly all chromosomes are degraded and most chromosome morphology is lost; and intermediate C-Frag, where extensive chromosomal degradation may have occurred but chromosome structure is largely still apparent

(Fig. 5.1) (Stevens et al. 2007). C-Frag can further be classified based on when it occurs during mitosis. C-Frag occurring during or following metaphase or after extended mitotic arrest results in small, tightly condensed chromosomes that are degraded during the process. C-Frag that occurs before metaphase results in degraded chromosomes that tend to be longer, not overly condensed, and have sister chromatids which have not separated (Fig. 5.2).

C-Frag Occurs as a Response to Stress

Mechanistically, C-Frag occurs as a general response to stress (Stevens et al. 2011a). C-Frag was originally identified in cells treated with genotoxic drugs; however, C-Frag is not caused directly by genotoxicity by a specific drug/reagent. C-Frag has been shown to occur in response to a number of treatments that can broadly be summed as stresses to the cellular system. Initially, C-Frag was described as chromosome pulverization resulting from measles infection, but it has subsequently also been shown to occur in knockouts of ATR, ATM, and p53 (Nichols and Levan 1965; Brown and Baltimore 2000; Fukasawa et al. 1997). Furthermore, C-Frag can be induced by endoplasmic reticulum stressors such as A31187, DTT, and thapsigargin. In fact, simply increasing the temperature at which cells are cultured increases the frequency of C-Frag. Finally, stress induced by the effects of genomic instability results in C-Frag. Cell lines with high levels of genomic instability also show increased levels of spontaneous C-Frag (Stevens et al. 2011a).

Further support for the mechanistic link between stress and C-Frag is evidenced by centrosome amplification (Stevens et al. 2011a). Centrosome amplification also occurs during times of stress. HSP90 is a core component of the centrosome, and inhibition of its function with 17-DMAG increases C-Frag (Stevens et al. 2011a). Taken together, C-Frag can be described as a mitotic cell death that occurs when mitotic cells are exposed to stress in general or which occurs when cells that have encountered lethal levels of stress are able to bypass cell-cycle checkpoints and enter mitosis (Fig. 5.4) (Stevens et al. 2011a).

C-Frag and Genomic Instability

C-Frag has direct implication for genomic instability. Cell-cycle checkpoint function is often abrogated in cells with genomic instability such as cancer cells. Loss of cell-cycle checkpoint function is common in cancer. Thus, mitotic cell death is the major type of cell death that occurs during cancer therapy, and C-Frag is a major form of mitotic cell death. Normally, damaged or highly stressed cells are arrested at cell-cycle checkpoints where the damage can then be fixed, or signals for apoptosis are given. In cells with genome change, however, gene networks are altered and their function changes, resulting in the ability to escape checkpoints;

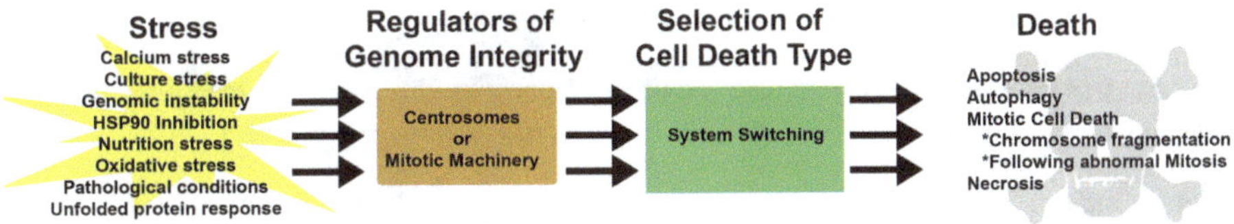

Fig. 5.4 Relationship between cellular stress and death. Cells subjected to levels of stress respond in multiple ways. Sublethal doses of stress in most cases result in adaptation of the cell without disruption to the genome system. Lethal doses of stress, however, affect individual cells differentially. The type of cell death that is induced is dependent on a number of factors including the integrity of the affected cell genome, the current state of the cell (such as the point in the cell cycle where the cell resides during the stress encounter or the intactness of cell-cycle checkpoints), the availability of cell death-inducing networks, and the degree and type of stress that is encountered (Adapted from Stevens et al. 2011a)

therefore, mitotic cell deaths, especially C-Frag, are the last line of defense to eliminate cells with altered genomes.

In the case of genomic instability, C-Frag can be a double-edged sword. Although C-Frag does eliminate cells with altered genomes, C-Frag can induce further genome change. C-Frag-related genomic change occurs in three ways. First, C-Frag can eliminate single (or multiple) chromosomes without inducing cell death, resulting in aneuploidy (Stevens et al. 2007). This chromosome elimination is more common when treatment dosages are sublethal. Interestingly, in yeast it has been shown that genome duplication followed by the loss of chromosomes is a major mechanism in the development of aneuploidy. C-Frag is one way to lose chromosomes and create aneuploidy. Second, in certain cases the process of C-Frag can stop, leaving large regions of chromosomes partially digested. Repair mechanisms such as nonhomologous end-joining can repair these partially digested chromosomes, resulting in highly rearranged chromosomes that are indicative of genome chaos. Interestingly, this could provide a mechanism for chromothripsis, one subtype of genome chaos where portions of one or more chromosomes are inserted into another chromosome, resulting in a highly rearranged chromosome with repeating segments of other chromosomes (Stephens et al. 2011; Liu et al. 2013). Third, a fraction of C-Frag cases directly contribute to genome chaos (Liu et al. 2013). Following induction of C-Frag, chromosome repair mechanisms including nonhomologous end-joining reattach chromosomal fragments to form new chimeric chromosomes (Fig. 5.5). Cells with genome chaos persist typically only for a few weeks until a stable genome is selected, but in some cases genome chaos can persist (Stephens et al. 2011; Liu et al. 2013).

Utility of C-Frag

The implications and utility of C-Frag are wide ranging. First, C-Frag functions in concert with other forms of cell death to eliminate diseased, stressed, or altered

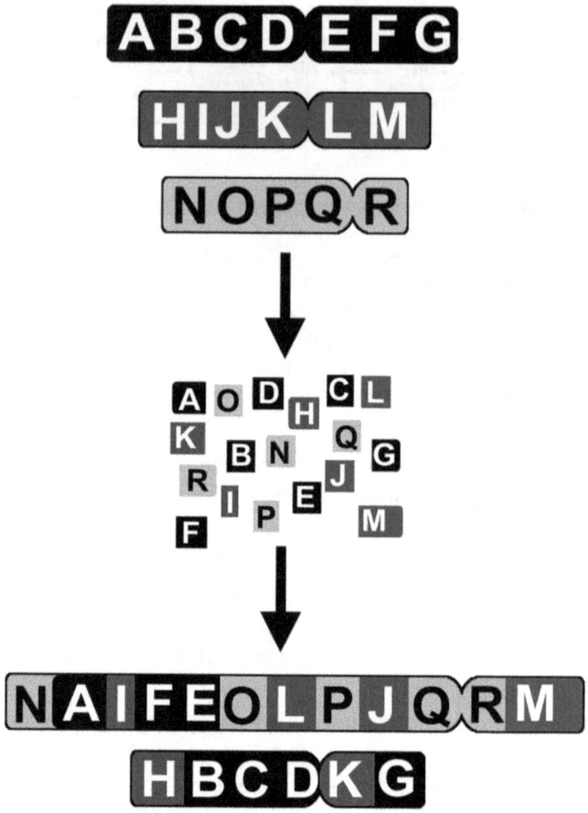

Fig. 5.5 Incomplete C-Frag can lead to multiple types of genome chaos. Following exposure to stress, normal chromosomes (*top*) undergo C-Frag (*middle*). Activation of various repair mechanisms such as nonhomologous end-joining results in rejoining of the fragments in a manner that reshuffles the genome (represented by the changing of the lettered regions of the genome and the chromosome number), resulting in genome chaos (*bottom*) (Adapted from Heng et al. 2011b)

cells (Stevens et al. 2011a). In combination with measures of other types of cell death, C-Frag is useful in measuring the amount of cell death occurring in a given sample. C-Frag is especially useful in measuring cell death in cancer samples because mitotic cell deaths are the main types of cell death that occur during cancer treatment and C-Frag is a major form of mitotic cell death. Therefore, C-Frag has obvious utility in evaluating the efficacy of new chemotherapeutic drugs. Second, C-Frag is a type of nonclonal chromosome aberration (NCCA) and is useful as an indicator of genome stability (Stevens et al. 2011a). Spontaneous C-Frag increases with genome instability. Thus, the presence of C-Frag occurring spontaneously at increased frequencies indicates increased genome instability. Incomplete C-Frag can lead to further induction of genomic instability through the development of genome chaos or aneuploidy. Therefore, C-Frag also gives a window into the probability of genome change, which in turn can be used to monitor the evolutionary potential for targeted cells (Stevens et al. 2007, 2011a, b, 2013b).

Premature Chromosome Condensation (PCC)

Definition and Historical Perspectives

Premature chromosome condensation (PCC) is a process whereby interphase chromatin is induced to condense into chromosomes abnormally (Johnson and Rao 1970; Sperling and Rao 1974a; Rao 1982). This condensation should not be confused with apoptotic condensation where degraded interphase DNA condenses (Figs. 5.3 and 5.6) (Martelli et al. 1997). During PCC, mitosis-promoting factor (MPF), which is composed of cyclin b and CDK1 (cdc2p34), is activated, moves from the cytoplasm to the nucleus, stimulating the events of mitosis, including breakdown of the nuclear envelope and condensation of chromatin (Nurse et al. 1976; Masui 2001; Wasserman and Masui 1976). The morphology of PCC differs depending on the stage of interphase in which PCC is induced. Induction of PCC during G_1 or G_2 results in condensation of complete chromosomes (Fig. 5.6) (Johnson and Rao 1970; Potu et al. 1977; Hanks et al. 1983). In the case of G_1 PCC, the chromatin has not been replicated, so the

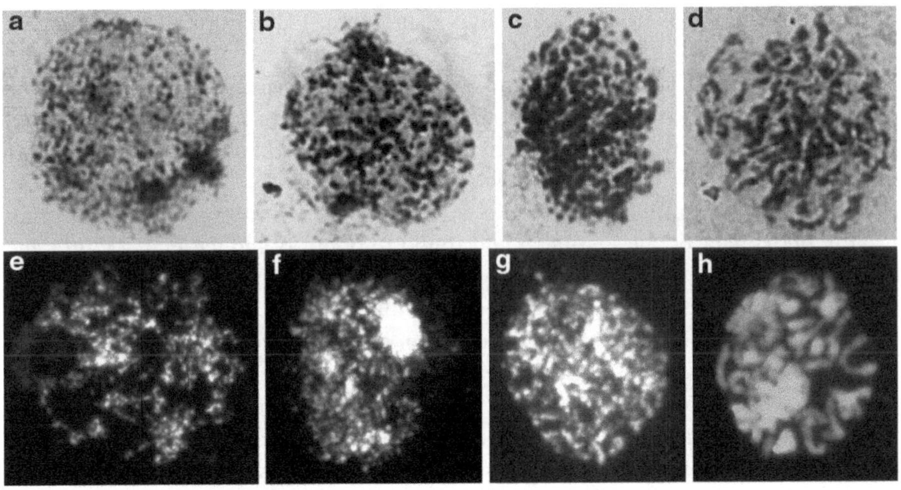

Fig. 5.6 Examples of various stages of premature chromosome condensation (PCC): Giemsa (**a**–**d**) and DAPI (**e**–**h**) staining. Treatment of cells with the phosphatase inhibitor calyculin A induces PCC in cells where cyclin b and CDK1 are present, primarily S-phase and G_2. The morphology of the resultant PCC figure depends on the stage of the cell cycle in which PCC is induced. During S-phase PCCs take on a "pulverized" appearance (**a**–**c**, **e**–**g**). PCCs in early S-phase cells result in condensation of small regions of the chromosomes (**a**, **e**). The visibly condensed regions have replicated whereas the unreplicated regions remain less tightly condensed. As S-phase proceeds, larger portions of the genome are replicated and chromosome morphology becomes more apparent (**b**, **c**; **f**, **g**). Notice the uniformity in size of the condensed regions within each cell. G_2 PCC causes condensation of fully replicated chromosomes, resulting in long chromosomes with little chromatid separation (**d**, **h**)

chromosomes lack sister chromatids. G_2 PCC chromosomes appear similar to early prophase chromosomes: they tend to be long, skinny, and not very condensed. Induction of PCC during S-phase results in the most unique chromosome morphology of the three stages of interphase. S-phase chromatin has not been fully replicated, and PCC induced during S-phase results in figures where replicated regions of chromosomes have normal condensed chromosome morphology while intervening unreplicated regions appear to lack chromatin (Fig. 5.6). This appearance can lead to the false conclusion that there are breaks between the replicated regions.

PCC was first described in 1970 when it was found that fusion of mitotic and interphase cells by the Sendai virus would cause condensation of the chromatin from the interphase cell (Johnson and Rao 1970). This finding demonstrated that mitotic cells could induce condensation of interphase chromatin but did not identify the reason for this. Subsequent work showed that MPF was the factor that initiated the PCC process (Masui and Markert 1971). MPF was soon after identified as a heterodimer of cyclin b and cdc2 (CDK1) (Lee and Nurse 1987).

Other methods to induce PCC have also been developed (Miura and Blakely 2011). Soon after the discovery that inactive Sendai virus could induce fusion-based PCC, protocols were developed that used ethylene glycol to induce fusion, eliminating the need for viral production (Lau et al. 1977). A number of chemicals have also been discovered to be capable of inducing PCC. For example, caffeine in combination with inhibitors of DNA synthesis such as hydroxy urea has been shown to induce PCC (Rybaczek and Kowalewicz-Kulbat 2011). The maturation of MPF requires phosphorylation of CDK1, which occurs spontaneously without dephosphorylation by protein phosphatase 2C (PP2C). Okadaic acid was found to dephosphorylate PP2C, inactivating it and inducing PCC (Ghosh et al. 1992). Subsequently calyculin A, a molecule with increased affinity for PP2C over other protein phosphatases, was isolated and found to potently induce PCC (Miura and Blakely 2011).

PCC has also been shown to occur as a result of various genetic manipulations. Attenuation of the G_2 checkpoint by deletion of key checkpoint proteins such as ATR can sensitize cells to PCC, especially in the presence of inhibitors of DNA synthesis (Nghiem et al. 2001). Other systems have been developed where PCC is induced in syncytia by coculture of one cell line expressing a receptor such as CD4 and another expressing a ligand such as the HIV envelope protein (Castedo et al. 2002, 2004c). As further discussed next, many phenomena previously described as PCC are actually C-Frag. Such confusion is caused by similar phenotypical characteristics.

Induction of PCC by fusion or by inhibitors of PP2C does not differ mechanistically because both types of PCC are caused by the exposure of interphase chromatin to activated MPF, regardless of the origin of the MPF. Aside from fusion-based PCC resulting in two closely located mitotic figures (the original mitotic chromosomes and the PCCs), G_1 PCC is rarely detectable by drug-induced PCC. Drug-induced PCC requires the presence of cyclin b and CDK1 to have an effect, and cyclin b in most cases is not expressed until S-phase (Bezrookove et al. 2003).

Application of PCC

The major application of PCC has been to induce condensation of interphase chromatin to allow for cytogenetic analysis (Bezrookove et al. 2003; Potu et al. 1977; Cheng et al. 1993). In addition, the PCC method has been used to address many basic questions of cell biology, and it has been instrumental in identification of the proteins and complexes of the cell cycle, especially mitosis (Potu et al. 1977; Sperling and Rao 1974b). PCC has also aided in developing the current understanding of how mitosis occurs. More recently PCC has been used to uncover unique mechanisms of DNA repair and to determine bystander effects of nuclei exposed to radiation (Terzoudi et al. 2008, 2010). In addition to its use in basic research, PCC has been an important method for karyotyping tissues and cells with low mitotic indices, as mitotic figures are required for cytogenetic analysis (Bezrookove et al. 2003; Miura and Blakely 2011). PCC is commonly used to measure radiation exposure and DNA damage caused by other exposures (Miura and Blakely 2011; Balakrishnan et al. 2010). Interestingly, fusion-based PCC could be used to determine the stage of the cell cycle that is the most susceptible to various forms of DNA damage. In this method PCC is induced in lymphocytes from patients with radiation exposure to visualize chromosome breaks to ascertain radiation exposures. PCC has also been used to karyotype tumor tissue as it reduces the need for cell culture. It is important to optimize conditions of PCC induction when it is used for monitoring DNA damage, because the condition of the cells in which PCC is induced, the concentration of okadaic acid or calyculin A, and the time of treatment all affect the quality of the figures produced (Miura and Blakely 2011).

Direct Comparison of C-Frag and PCC

Differences

C-Frag and S-phase PCC figures can be difficult to differentiate, but the phenomena can be differentiated by a number of factors (Table 5.1). Fusion-based PCC is the easiest form of PCC to differentiate from C-Frag as fusion-based PCC results in figures containing a mitotic cell located closely to the nucleus undergoing PCC. Although there are instances of the two figures overlapping, this is uncommon. Visual assessment of ploidy levels of the involved nuclei can be helpful in cases of overlap. Although most cell lines have some degree of aneuploidy, nuclei from fused cells can be differentiated from aneuploidy cells by having an idea of the average chromosome count of the studied population. Differentiation of nonfusion-based PCC and C-Frag is slightly more difficult. Morphologically, C-Frag results in a higher diversity of chromosomal morphology within a given mitotic figure. C-Frag, especially in cases of intermediate-stage fragmentation, produces mitotic

Table 5.1 Diverse mechanisms of chromosome fragmentation (C-Frag)

Factors of pulverization/shattering	Species	Year	References
UV exposure	*Tradescantia paludosa*	1954	Lovelace (1954)
Exposure of male mice to methyl methane-sulfonate before fertilization of female mice. Shattering seen in filial cells	Mouse	1975	Brewen et al. (1970)
Infection of lymphocytes with JM-V herpesvirus	Chicken	1976	Yoon et al. (1976)
Treatment with fungicide *N*-trichloromethylthio-phthalimide	Human	1978	Sirianni and Huang (1978)
Pulverization caused by UV light and caffeine	Chinese hamster	1980	Cremer et al. (1980)
Exposure to alpha-amanitin	Rat	1985	Magalhães and Magalhães (1985)
Doxorubicin treatment. Pulverization inhibited in drug-resistant cells	Human	1986	Tapiero et al. (1986)
Herpes simplex virus type 1 infection	Human	1986	Peat and Stanley (1986)
Hepatitis B infection. Pulverization occurs in both hepatocellular cell line derived from a tumor and in peripheral lymphocytes from the patient	Human	1986	Simon and Knowles (1986)
Herpes simplex virus (HSV) type 1 infection. Endoreduplication was noted. Also HSV infection increased the mitotic index		1986	Chenet-Monte et al. (1986)
Friend leukemia cells exposed to high levels of ADM	Mouse	1986	Patet et al. (1986)
Following incubation of cells with temperature-sensitive DNA polymerase α in S-phase at 39 °C that were then cultured in a permissive temperature	Mouse	1987	Eki et al. (1987)
Photoirradiation of G_2 or early prophase cells	CHO	1990	Fernandez et al. (1990)
Hepatitis B integration and genomic instability	Human	1993	Grabovskaya et al. (1993)
N-Methyl-*N*′-nitro-*N*-nitrosoguanidine (MNNG), sodium selenite, and caffeine treatment of CHO cells	Chinese hamster	1994	Balansky et al. (1994)
Culture of a fibroblast line generated from a patient with xeroderma	Human	1995	Casati et al. (1995)
Associated with ubiquitin-activating enzyme E1 activity	Mouse	1995	Sudha et al. (1995)
Vaccination against hog cholera virus	Swine	1998	Genghini et al. (1998)
Radiation exposure	Mouse	1998	Ganasoundari et al. (1998)
Maintenance of diploid karyotype in PA-1 cells by removal of tetraploid cells	Human	1999	Gao et al. (1999)

(continued)

Table 5.1 (continued)

2-Methoxyestradiol, an endogenous metabolite of estrogen	Human	2000	Tsutsui et al. (2000)
Following vaccination of pigs for swine fever	Swine	2002	Genghini et al. (2002)
Streptozotocin treatment	Human	2003	Bolzán and Bianchi (2003)
Viral infection in pigs	Swine	2004	Sutiaková et al. (2004)
Vitamin C treatment of lymphocytes	Human	2008	Nefíc (2008)
Shattering by UV light and caffeine exposure	Chinese hamster	2009	Hübner et al. (2009)

Source: Adapted from Stevens et al. (2010)

figures where some chromosomes are degraded to the point that chromosomal morphology is lost while some chromosomes may remain intact. In contrast, the chromosomal morphology and the number of gaps between condensed regions in S-phase PCC chromosomes is dependent on the degree of replication that has occurred. Although certain chromosomes may replicate at different times, overall replication tends to occur similarly across the genome. Thus, induction of PCC in S-phase cells results in figures with a more regular morphology than those produced by C-Frag. An exception to this rule occurs in micronucleated cells. Micronuclei can replicate at different times than does the main nucleus. When the cell completes G_2 and enters into M-phase, partially replicated micronuclei can be induced to undergo PCC, resulting in a small region of a mitotic figure showing similar morphology to C-Frag. Care should be taken during cytogenetic analysis to note the prevalence of micro- and multinucleated cells within the population as C-Frag may be over estimated in these populations. C-Frag occurring in later stages of mitosis can also be easily differentiated as PCC does not produce highly condensed chromosomes. Taken together, careful morphological inspection during cytogenetic analyses can differentiate C-Frag and PCC.

C-Frag and PCC can also be differentiated biochemically. Although morphologically S-phase PCC figures appear to have breaks interspersed in regions of condensed chromatin, these gaps are not formed by DNA breaks (Gollin et al. 1984): they are simply composed of the unreplicated regions of DNA that do not readily condense. The lack of DNA breaks in S-phase PCC has been confirmed by electron microscopy and the lack of γ-H2AX staining, which is indicative of double-strand breaks. The chromosomal pieces observed in C-Frag, on the other hand, show extensive γ-H2AX staining, although to date electron microscopy has not been performed on C-Frag. A prepulse of bromodeoxyuridine (5-bromo-2'-deoxyuridine, BrdUrd) before induction of S-phase PCC demonstrates active DNA replication during S-phase PCC whereas culturing mitosis-arrested cells in BrdUrd and subsequently inducing C-Frag by doxorubicin treatment did not reveal active DNA replication, demonstrating that C-Frag occurs during mitosis and is not related to S-phase. Furthermore, during C-Frag induction of PCC does not result in PARP degradation. Outside the nucleus, PCC and C-Frag also differ. C-Frag is associated with multiple centrosomes while spindle formation is inhibited during PCC (Ghosh et al. 1992). Thus, C-Frag and PCC are distinguishable at both morphological and biochemical levels.

Overlap

Although PCC and C-Frag can be differentiated in most cases, there are some areas of overlap between the two phenomena. This overlap centers on the G_2 checkpoint. In cancer, the checkpoint function is often abrogated (Kastan and Bartek 2004), allowing cells with DNA damage that would normally arrest before mitosis until that damage was repaired to abnormally enter mitosis. Entry into mitosis with DNA damage then can result in the induction of C-Frag (Stevens et al. 2011a). Thus, cells that abnormally slip past the G_2 checkpoint are prematurely entering mitosis. Although C-Frag and PCC are distinct phenomena, as with many cellular phenomena there is some overlap.

Pulverization, Shattering, and Mitotic Catastrophe

Pulverization and Shattering

A number of reports of chromosome pulverization and chromosome shattering predate discovery of PCC (Lovelace 1954; Nichols and Levan 1965; Kato and Sandberg 1967). Subsequent discovery of PCC led to the shattered figures being called PCC, although there must be some doubt to PCC being the mechanism in all cases of shattering or pulverization because the continued use of the terms 'shattering' and 'pulverization'. Although some cases of pulverization are indeed PCC, such as in the case of binucleate cells in which the nuclei asynchronously replicate and in cases of cellular fusion that result in obvious PCC figures (Kato and Sandberg 1967), other cases wherein pulverization is induced by various widely ranging stresses, including viral infection, chemical exposure, radiation exposure, and genetic defects, this pulverization is most likely C-Frag (Table 5.2) (Stevens et al. 2010; Nichols and Levan 1965). Recently chromosome pulverization has been raised as a potential mechanism behind chromothripsis, a form of genome chaos where a single chromosome recombines multiple times to form a highly complex chromosome (Crasta et al. 2012). However, pulverization is synonymous with PCC, and PCC does not result in strand breaks; thus, pulverization cannot contribute to chromothripsis (Stevens et al. 2010; Micronuclear chromosome pulverization may underlie chromothripsis 2012; Gollin et al. 1984). Chromosome pulverization, or shattering, is a morphological description, whereas C-Frag and PCC are both morphological and mechanistic descriptions. Therefore, the use of chromosome pulverization/shattering should be avoided in favor of either S-phase PCC or C-Frag, depending on which process is observed. As we previously suggested, the use of C-Frag will reduce such confusion (Stevens et al. 2010).

Table 5.2 Identifiable differences of chromosome fragmentation and premature chromosome condensation (PCC)

Chromosome fragmentation	PCC
Morphological	
Single cell involvement	If fusion induced, normal, intact mitotic cells will be in close proximity to fragmented cells
Can affect single chromosomes	Impacts entire genome regularly, except in limited multinucleated cells
Results in chromosome degradation	Unknown, may activate chromosome breakdown
Chromosome morphology lost as process progresses	Chromosome morphology dependent on position in cell cycle
Differential cut size	Differential condensation states
Mechanistic	
Occurs during mitosis	Occurs in interphase cells exposed to active MPF
Not inhibited by roscovitine	Inhibited by roscovitine
Induced by stress during mitosis	Induced by cell fusion or activation of MPF
γ-H2AX positive	γ-H2AX negative
No active DNA incorporation	Actively incorporating DNA

Source: Previously published from Stevens et al. (2010)

Mitotic Catastrophe

The term mitotic catastrophe has been applied to a number of phenomena where cell death is linked to mitosis. Various reports have shown mitotic catastrophe to occur with multiple phenotypes, which range from cell death following abnormal mitosis where apoptotic pathways are activated, to cell deaths that occur directly during mitosis which may or may not be associated with apoptosis (Chan et al. 1999; Roninson et al. 2001; Castedo et al. 2004b; Hübner et al. 2009). Despite warnings against its use because of the lack of a solid, recurrent morphological or mechanistic definition, the term mitotic catastrophe has remained pervasive in the literature (Kroemer et al. 2005). Many reports of mitotic catastrophe rely on DNA content measures to define mitotic cells without using specific markers of mitosis such as histone H3 phosphorylation at serine 10 or without cytogenetic or in-depth morphological characterization. This carelessness has led to increased confusion in the field.

C-Frag has been shown to differ from a well-described model of mitotic catastrophe. In this model, cells lacking 14-3-3σ, which plays a role in the G_2 checkpoint, have been shown to undergo a mitotic cell death following low-dose doxorubicin treatment; however, when these cells were treated no C-Frag was detectable (Stevens et al. 2007; Chan et al. 1999). Other models of mitotic catastrophe have been developed that are based on cellular fusion (Castedo et al. 2002, 2004a). In this case it is likely that a significant portion of cells are undergoing PCC. The fact that descriptions of mitotic catastrophe are wide ranging should, however, not detract from the important message that mitotic cell deaths play a major role in the reduction of tumor size following chemotherapy. It is apparent that at least two distinct types of mitotic cell death occur: C-Frag, which occurs directly during mitosis, and another form of cell death that occurs following an abnormal or failed mitosis.

Future Perspectives

A major drawback of cytogenetic analyses is the requirement for condensed chromosomes to perform most analyses. PCC has proven to be an important cytogenetic method that increases the ability to observe mitotic figures in cells or tissues which are not actively dividing, and it will continue to have multiple clinical and basic research implications. PCC has proven to be instrumental in the development of our knowledge of the cell cycle, especially the proteins and complexes that are involved in mitosis. PCC will continue to be an important tool in measuring DNA breaks associated with radiation and other hazardous exposures. Development of improved methods of inducing PCC, especially methods of inducing G_1 PCC to allow cytogenomic analysis of tissues with little or no mitotic activity, will improve the power of PCC and cytogenomics in general.

Multiple questions about C-Frag remain and should be addressed in future work. First, although C-Frag differs from apoptosis, these two types of cell death appear to complement each other in that they both eliminate unfit cells. Further work will determine potential links between these deaths and other deaths such as autophagy and necrosis, and whether other forms of cell death may be activated late in the process of C-Frag (Stevens et al. 2007). Determination of whether there is a specific order of C-Frag may reveal more of the biology behind C-Frag. Identification of whether there are specific sequence motifs that are targeted early in C-Frag, and whether portions of the genome are more resistant to degradation, would also be interesting; current genomic technologies such as next-generation sequencing put this research within grasp and could aid in identification of the proteins/systems directly responsible for fragmenting the chromosomes.

As C-Frag represents a type of NCCA, a key question is what is the relationship between C-Frag and overall frequency of the NCCAs? How to use the C-Frag to predict the outcome of cancer therapy and potential risk of induced genome chaos? Increasing the knowledge of how C-Frag is related to genome chaos and other chromosome aberrations may provide new insights into the somatic evolution leading to cancer. Finally, as C-Frag may also be involved in developmental processes, more research is needed to address its significance in both development and evolution (Fujiwara et al. 1997; Gernand et al. 2005).

Conclusion

Chromosome fragmentation and PCC are processes that are distinct both morphologically and mechanistically, although they have been confused in the past. C-Frag is a major form of mitotic cell death that is induced by various forms of stress whereas PCC is a phenomenon in which interphase chromatin is forced to abnormally condense outside mitosis. Both C-Frag and PCC are relevant for both basic research and in the medical clinic, and future studies of both phenomena will increase our understanding of the chromatin and chromosomes.

References

Alam MT, Kasatiya S (1976) Cytological effects of an organic phosphate pesticide on human cells in vitro. Can J Genet Cytol 18(4):665–671

Balakrishnan S, Shirsath K, Bhat N, Anjaria K (2010) Biodosimetry for high dose accidental exposures by drug induced premature chromosome condensation (PCC) assay. Mutat Res 699(1–2):11–16. doi:10.1016/j.mrgentox.2010.03.008

Balansky R, Adomat H, Bryson L (1994) Additive coclastogenicity of sodium selenite and caffeine in CHO cells treated with *N*-methyl-*N*′-nitro-*N*-nitrosoguanidine. Biol Trace Elem Res 42(1):53–61

Bezrookove V, Smits R, Moeslein G, Fodde R, Tanke HJ, Raap AK, Darroudi F (2003) Premature chromosome condensation revisited: a novel chemical approach permits efficient cytogenetic analysis of cancers. Genes Chromosomes Cancer 38(2):177–186

Bolzán AD, Bianchi MS (2003) Clastogenic effects of streptozotocin on human colon cancer cell lines with gene amplification. J Environ Pathol Toxicol Oncol 22(4):281–286

Brewen JG, Nettesheim P, Jones KP (1970) A host-mediated assay for cytogenetic mutagenesis: preliminary data on the effect of methyl methanesulfonate. Mutat Res 10(6):645–649

Brown EJ, Baltimore D (2000) ATR disruption leads to chromosomal fragmentation and early embryonic lethality. Genes Dev 14(4):397–402

Casati A, Riboni R, Caprioli J, Nuzzo F, Mondello C (1995) Condensation anomalies and exclusion in micronuclei of rearranged chromosomes in human fibroblasts cultured in vitro. Chromosoma (Berl) 104(2):137–142

Castedo M, Perfettini JL, Roumier T, Kroemer G (2002) Cyclin-dependent kinase-1: linking apoptosis to cell cycle and mitotic catastrophe. Cell Death Differ 9(12):1287–1293

Castedo M, Perfettini JL, Roumier T, Valent A, Raslova H, Yakushijin K, Horne D, Feunteun J, Lenoir G, Medema R, Vainchenker W, Kroemer G (2004a) Mitotic catastrophe constitutes a special case of apoptosis whose suppression entails aneuploidy. Oncogene 23(25):4362–4370

Castedo M, Perfettini JL, Roumie T, Andreau K, Medema R, Kroemer G (2004b) Cell death by mitotic catastrophe: a molecular definition. Oncogene 23(16):2825–2837

Castedo M, Perfettini JL, Roumier T, Yakushijin K, Horne D, Medema R, Kroemer G (2004c) The cell cycle checkpoint kinase Chk2 is a negative regulator of mitotic catastrophe. Oncogene 23(25):4353–4361

Chan TA, Hermeking H, Lengauer C, Kinzler KW, Vogelstein B (1999) 14-3-3σ is required to prevent mitotic catastrophe after DNA damage. Nature (Lond) 401(6753):616–620

Chenet-Monte C, Mohammad F, Celluzzi CM, Schaffer PA, Farber FE (1986) Herpes simplex virus gene products involved in the induction of chromosomal aberrations. Virus Res 6(3):245–260

Cheng X, Pantelias GE, Okayasu R, Cheong N, Iliakis G (1993) Mitosis-promoting factor activity of inducer mitotic cells may affect radiation yield of interphase chromosome breaks in the premature chromosome condensation assay. Cancer Res 53(23):5592–5596

Crasta K, Ganem NJ, Dagher R, Lantermann AB, Ivanova EV, Pan Y, Nezi L, Protopopov A, Chowdhury D, Pellman D (2012) DNA breaks and chromosome pulverization from errors in mitosis. Nature (Lond) 482(7383):53–58. doi:10.1038/nature10802

Cremer C, Cremer T, Simickova M (1980) Induction of chromosome shattering and micronuclei by ultraviolet light and caffeine. I. Temporal relationship and antagonistic effects of the four deoxyribonucleosides. Environ Mutagen 2(3):339–351

Eki T, Enomoto T, Murakami Y, Miyazawa H, Hanaoka F, Yamada M (1987) Characterization of revertants derived from a mouse DNA temperature-sensitive mutant strain, tsFT20, which contains heat-labile DNA polymerase alpha activity. Exp Cell Res 171(1):24–36

Fernandez JL, Costas E, Goyanes VJ (1990) Chromosome structure and condensation in relation to DNA integrity. Cytobios 63(254–255):193–204

Fujiwara A, Abe S, Yamaha E, Yamazaki F, Yoshida MC (1997) Uniparental chromosome elimination in the early embryogenesis of the inviable salmonid hybrids between masu salmon female and rainbow trout male. Chromosoma (Berl) 106(1):44–52

Fukasawa K, Wiener F, VandeWoude GF, Mai SB (1997) Genomic instability and apoptosis are frequent in p53 deficient young mice. Oncogene 15(11):1295–1302

Ganasoundari A, Devi PU, Rao BS (1998) Enhancement of bone marrow radioprotection and reduction of WR-2721 toxicity by *Ocimum sanctum*. Mutat Res 397(2):303–312

Gao C, Miyazaki M, Ohashi R, Tsuji T, Inoue Y, Namba M (1999) Maintenance of near-diploid karyotype of PA-1 human ovarian teratocarcinoma cells due to death of polyploid cells by chromosome fragmentation/pulverization. Int J Mol Med 4(3):291–294

Genghini R, Tiranti I, Segade G, Amado J, Wittouck P, Mian L (1998) In vivo effect on pig chromosomes of high dosage vaccine against classic swine fever. Mutat Res 422(2):357–365

Genghini R, Tiranti I, Wittouck P (2002) Pig chromosome aberrations after vaccination against classical swine fever in field trials. Vaccine 20(23–24):2873–2877

Gernand D, Rutten T, Varshney A, Rubtsova M, Prodanovic S, Bruss C, Kumlehn J, Matzk F, Houben A (2005) Uniparental chromosome elimination at mitosis and interphase in wheat and pearl millet crosses involves micronucleus formation, progressive heterochromatinization, and DNA fragmentation. Plant Cell 17(9):2431–2438

Ghosh S, Paweletz N, Schroeter D (1992) Failure of kinetochore development and mitotic spindle formation in okadaic acid-induced premature mitosis in HeLa cells. Exp Cell Res 201(2):535–540

Gollin SM, Wray W, Hanks SK, Hittelman WN, Rao PN (1984) The ultrastructural organization of prematurely condensed chromosomes. J Cell Sci Suppl 1:203–221

Gorelick R, Heng HH (2011) Sex reduces genetic variation: a multidisciplinary review. Evolution 65(4):1088–1098. doi:10.1111/j.1558-5646.2010.01173.x

Grabovskaya IL, Tugizov SM, Glukhova LA, Kushch AA (1993) Cytogenetic analysis of human hepatocarcinoma cell line PLC-PRF-5 and its mutant clones with different degrees of cell differentiation. Cancer Genet Cytogenet 65(2):147–151

Hanks SK, Gollin SM, Rao PN, Wray W, Hittelman WN (1983) Cell cycle-specific changes in the ultrastructural organization of prematurely condensed chromosomes. Chromosoma (Berl) 88(5):333–342

Heng HH (2007) Elimination of altered karyotypes by sexual reproduction preserves species identity. Genome 50(5):517–524. doi:g07-039 pii10.1139/g07-039

Heng HH (2009) The genome-centric concept: resynthesis of evolutionary theory. Bioessays 31(5):512–525. doi:10.1002/bies.200800182

Heng HH (2013) 4-D genomics: genome dynamics and constraint in evolution. Springer, New York (in press)

Heng HH, Stevens JB, Liu G, Bremer SW, Ye CJ (2004) Imaging genome abnormalities in cancer research. Cell Chromosome 3(1):1. doi:10.1186/1475-9268-3-1

Heng HH, Liu G, Bremer S, Ye KJ, Stevens J, Ye CJ (2006) Clonal and non-clonal chromosome aberrations and genome variation and aberration. Genome 49(3):195–204

Heng HH, Bremer SW, Stevens JB, Ye KJ, Liu G, Ye CJ (2009) Genetic and epigenetic heterogeneity in cancer: a genome-centric perspective. J Cell Physiol 220(3):538–547. doi:10.1002/jcp.21799

Heng HH, Stevens JB, Bremer SW, Ye KJ, Liu G, Ye CJ (2010a) The evolutionary mechanism of cancer. J Cell Biochem 109(6):1072–1084

Heng HH, Liu G, Stevens JB, Bremer SW, Ye KJ, Ye CJ (2010b) Genetic and epigenetic heterogeneity in cancer: the ultimate challenge for drug therapy. Curr Drug Targets 11(10):1304–1316

Heng HH, Liu G, Stevens JB, Bremer SW, Ye KJ, Abdallah BY, Horne SD, Ye CJ (2011a) Decoding the genome beyond sequencing: the new phase of genomic research. Genomics 98(4):242–252. doi:10.1016/j.ygeno.2011.05.008

Heng HH, Stevens JB, Bremer SW, Liu G, Abdallah BY, Ye CJ (2011b) Evolutionary mechanisms and diversity in cancer. Adv Cancer Res 112:217–253

Heng HH, Bremer SW, Stevens JB, Horne SD, Liu G, Abdallah BY, Ye KJ, Ye CJ (2013a) Chromosomal instability (CIN): what is it and why is it crucial to cancer evolution? Cancer Met Rev (in Press)

Heng HH, Liu G, Stevens JB, Abdallah BY, Horne SD, Ye KJ, Bremer SW, Ye CJ (2013b) Karyotype heterogeneity and unclassified chromosomal abnormalities. Cytogenet Genome Res (in press)

Hübner B, Strickfaden H, Müller S, Cremer M, Cremer T (2009) Chromosome shattering: a mitotic catastrophe due to chromosome condensation failure. Eur Biophys J 38(6):729–747

Johnson RT, Rao PN (1970) Mammalian cell fusion: induction of premature chromosome condensation in interphase nuclei. Nature (Lond) 226(5247):717–722

Kastan M, Bartek J (2004) Cell-cycle checkpoints and cancer. Nature (Lond) 432:316–323

Kato H, Sandberg AA (1967) Chromosome pulverization in human binucleate cells following colcemid treatment. J Cell Biol 34(1):35–45

Kato H, Sandberg AA (1968) Chromosome pulverization in chinese hamster cells induced by Sendai virus. J Natl Cancer Inst 41(5):1117–1123

Kloosterman WP, Guryev V, van Roosmalen M, Duran KJ, de Bruijn E, Bakker SC, Letteboer T, van Nesselrooij B, Hochstenbach R, Poot M, Cuppen E (2011) Chromothripsis as a mechanism driving complex de novo structural rearrangements in the germline. Hum Mol Genet 20(10):1916–1924. doi:10.1093/hmg/ddr073

Knuutila S, Siimes M, Vuopio P (1981) Chromosome pulverization in blood diseases. Hereditas 95(1):15–24

Kroemer G, El-Deiry WS, Golstein P, Peter ME, Vaux D, Vandenabeele P, Zhivotovsky B, Blagosklonny MV, Malorni W, Knight RA, Piacentini M, Nagata S, Melino G (2005) Classification of cell death: recommendations of the Nomenclature Committee on Cell Death. Cell Death Differ 12(S2):1463–1467

Lau YF, Brown RL, Arrighi FE (1977) Induction of premature chromosome condensationin CHO cells fused with polyethlyene glycol. Exp Cell Res 110:57–61

Lee MG, Nurse P (1987) Complementation used to clone a human homologue of the fission yeast cell cycle control gene cdc2. Nature (Lond) 327(6117):31–35. doi:10.1038/327031a0

Liu G, Stevens JB, Abdallah BY, Horne S, Bremer SW, Ye CJ, Heng HH (2013) Genome chaos: survival strategy during crisis (submitted)

Lovelace R (1954) Chromosome shattering by ultraviolet radiation (2650 A). Proc Natl Acad Sci U S A 40(12):1129–1135

Magalhães MC, Magalhães MM (1985) Effects of alpha-amanitin on the fine structure of adrenal fasciculata cells in the young rat. Tissue Cell 17(1):27–37

Martelli AM, Bareggi R, Bortul R, Grill V, Narducci P, Zweyer M (1997) The nuclear matrix and apoptosis. Histochem Cell Biol 108(1):1–10

Masui Y (2001) From oocyte maturation to the in vitro cell cycle: the history of discoveries of maturation-promoting factor (MPF) and cytostatic factor (CSF). Differentiation 69(1):1–17. doi:10.1046/j.1432-0436.2001.690101.x

Masui Y, Markert CL (1971) Cytoplasmic control of nuclear behavior during meiotic maturation of frog oocytes. J Exp Zool 177(2):129–145. doi:10.1002/jez.1401770202

Micronuclear chromosome pulverization may underlie chromothripsis (2012). Cancer Discov 2(3):OF5. doi:10.1158/2159-8290.CD-RW2012-011

Miura T, Blakely WF (2011) Optimization of calyculin A-induced premature chromosome condensation assay for chromosome aberration studies. Cytometry A 79(12):1016–1022. doi:10.1002/cyto.a.21154

Nefíc H (2008) The genotoxicity of vitamin C in vitro. Bosn J Basic Med Sci 8(2):141–146

Nghiem P, Park PK, Kim Y, Vaziri C, Schreiber SL (2001) ATR inhibition selectively sensitizes G1 checkpoint-deficient cells to lethal premature chromatin condensation. Proc Natl Acad Sci U S A 98(16):9092–9097. doi:10.1073/pnas.161281798

Nichols WW, Levan A (1965) Measles associated chromosome breakage. Arch Gesamte Virusforsch 16:168–174

Norrby E, Levan A, Nichols WW (1966) The correlation between the chromosome pulverization effect and other biological activities of measles virus preparations. Exp Cell Res 41(3):483–491

Nurse P, Thuriaux P, Nasmyth K (1976) Genetic control of the cell division cycle in the fission yeast *Schizosaccharomyces pombe*. Mol Gen Genet 146(2):167–178

Patet J, Huppert J, Fourcade A, Tapiero H (1986) Cytogenetic modifications of Friend leukemia cells resistant to adriamycin. Leuk Res 10(6):651–658

Peat DS, Stanley MA (1986) Chromosome damage induced by herpes simplex virus type 1 in early infection. J Gen Virol 67(10):2273–2277. doi:10.1099/0022-1317-67-10-2273

Potu N, Rao BW, Puck TT (1977) Premature chromosome condensation and cell cycle analysis. J Cell Physiol 91(1):131–141

Rao P (1982) The phenomenon of premature chromosome condensation. In: Rao P, Johnson R, Sperling K (eds) Premature chromosome condensation application in basic, clinical and mutation research. Academic, New York, pp 2–37

Roninson IB, Broude EV, Chang BD (2001) If not apoptosis, then what?—Treatment-induced senescence and mitotic catastrophe in tumor cells. Drug Resist Updat 4(5):303–313

Rybaczek D, Kowalewicz-Kulbat M (2011) Premature chromosome condensation induced by caffeine, 2-aminopurine, staurosporine and sodium metavanadate in S-phase arrested HeLa cells is associated with a decrease in Chk1 phosphorylation, formation of phospho-H2AX and minor cytoskeletal rearrangements. Histochem Cell Biol 135(3):263–280. doi:10.1007/s00418-011-0793-3

Sandberg AA (1978) Some comments regarding chromosome pulverization (premature chromosome condensation or PCC, prophasing). Virchows Arch B Cell Pathol 29(1–2):15–18

Simon D, Knowles BB (1986) Hepatocellular carcinoma cell line and peripheral blood lymphocytes from the same patient contain common chromosomal alterations. Lab Invest 55(6): 657–665

Sirianni SR, Huang CC (1978) Effect of fungicide Folpet on growth and chromosomes of human lymphoid cell lines. Can J Genet Cytol 20(2):193–197

Sperling K, Rao PN (1974a) The phenomenon of premature chromosome condensation: its relevance to basic and applied research. Humangenetik 23(4):235–258

Sperling K, Rao PN (1974b) Mammalian cell fusion. V. Replication behaviour of heterochromatin as observed by premature chromosome condensation. Chromosoma (Berl) 45(2):121–131

Stephens PJ, Greenman CD, Fu B, Yang F, Bignell GR, Mudie LJ, Pleasance ED, Lau KW, Beare D, Stebbings LA, McLaren S, Lin ML, McBride DJ, Varela I, Nik-Zainal S, Leroy C, Jia M, Menzies A, Butler AP, Teague JW, Quail MA, Burton J, Swerdlow H, Carter NP, Morsberger LA, Iacobuzio-Donahue C, Follows GA, Green AR, Flanagan AM, Stratton MR, Futreal PA, Campbell PJ (2011) Massive genomic rearrangement acquired in a single catastrophic event during cancer development. Cell 144(1):27–40. doi:10.1016/j.cell.2010.11.055

Stevens JB, Liu G, Bremer SW, Ye KJ, Xu W, Xu J, Sun Y, Wu GS, Savasan S, Krawetz SA, Ye CJ, Heng HH (2007) Mitotic cell death by chromosome fragmentation. Cancer Res 67(16):7686–7694. doi:10.1158/0008-5472.CAN-07-0472

Stevens JB, Abdallah BY, Regan SM, Liu G, Bremer SW, Ye CJ, Heng HH (2010) Comparison of mitotic cell death by chromosome fragmentation to premature chromosome condensation. Mol Cytogenet 3:20

Stevens JB, Abdallah B, Liu G, Ye CJ, Horne SD, Wang G, Savasan S, Shekhar M, Krawetz SA, Huttemann M, Tainsky MA, Wu GS, Xie Y, Zhang K, Heng HH (2011a) Diverse system stresses: common mechanisms of chromosome fragmentation. Cell Death Dis 2:e178–e186

Stevens JB, Abdallah BY, Horne SD, Liu G, Bremer SW, Heng HH (2011b) Genetic and epigenetic heterogeneity in cancer. Encyc Life Sci doi:10.1002/9780470015902.a0023592

Stevens JB, Abdallah BY, Horne S, Bremer SW, Liu G, Ye CJ, Heng HH (2013a) Unstable genomes elevate transcriptome dynamics (submitted)

Stevens JB, Horne SD, Abdallah BY, Ye CJ, Heng HH (2013a) Chromosomal instability and transcriptome dynamics in cancer. Cancer Met Rev (in Press)

Stevens JB, Abdallah BY, Liu G, Horne SD, Bremer SW, Ye KJ, Huang JY, Kurkinen M, Ye CJ, Heng HH (2013b) Cell death heterogeneity. Cytogenet Genome Res (in press)

Sudha T, Tsuji H, Sameshima M, Matsuda Y, Kaneda S, Nagai Y, Yamao F, Seno T (1995) Abnormal integrity of the nucleolus associated with cell cycle arrest owing to the temperature-sensitive ubiquitin-activating enzyme E1. Chromosome Res 3(2):115–123

Sutiaková I, Rimková S, Sutiak V, Porácová J, Krajnicáková M, Harichová D (2004) A possible relationship between viral infection and chromosome damage in breeding boars. Berl Munch Tierarztl Wochenschr 117(1–2):16–18

Tapiero H, Patet J, Fourcade A, Huppert J (1986) Chromosomal changes associated with resistance to doxorubicin: correlation with tumorigenicity. Anticancer Res 6(2):203–208

Terzoudi GI, Singh SK, Pantelias GE, Iliakis G (2008) Premature chromosome condensation reveals DNA-PK independent pathways of chromosome break repair. Int J Oncol 33(4): 871–879

Terzoudi GI, Donta-Bakoyianni C, Iliakis G, Pantelias GE (2010) Investigation of bystander effects in hybrid cells by means of cell fusion and premature chromosome condensation induction. Radiat Res 173(6):789–801. doi:10.1667/RR2023.1

Tsutsui T, Tamura Y, Hagiwara M, Miyachi T, Hikiba H, Kubo C, Barrett JC (2000) Induction of mammalian cell transformation and genotoxicity by 2-methoxyestradiol, an endogenous metabolite of estrogen. Carcinogenesis (Oxf) 21(4):735–740

Wasserman WJ, Masui Y (1976) A cytoplasmic factor promoting oocyte maturation: its extraction and preliminary characterization. Science 191(4233):1266–1268

Ye CJ, Liu G, Bremer SW, Heng HH (2007) The dynamics of cancer chromosomes and genomes. Cytogenet Genome Res 118(2–4):237–246

Yoon JW, Kim SN, Hahn EC, Kenyon AJ (1976) Lymphoproliferative diseases of fowl: chromosome breaks caused in lymphocytes by JM-V herpesvirus. J Natl Cancer Inst 56(4):757–762

Chapter 6
Chromosomes and Nuclear Organization in ICF Syndrome

Emanuela V. Volpi

Abstract The ICF syndrome is a rare, autosomal recessive disorder, often fatal in childhood, and characterized by genetic and clinical heterogeneity. Its most consistent features are reduction in serum immunoglobulin levels, facial anomalies, and cytogenetic defects. ICF is also characterized by abnormal DNA methylation. Significant DNA hypomethylation is present mainly in the classical satellite sequences, the major constituent of the juxtacentromeric heterochromatin of chromosomes 1, 9, and 16. The relationship between DNA methylation defects, altered gene expression, and clinical and phenotypic features in ICF has been the object of intense scrutiny. Although the full pathogenetic picture remains to be elucidated, a number of hypotheses advocating an epigenetic model for this syndrome have been advanced by different research groups. Central to some of these hypotheses is the postulation of a trans-acting regulatory role for the heterochromatin and the suggestion of a possible connection between altered gene expression in ICF and the inappropriate release or recruitment of regulatory complexes by the hypomethylated satellite DNA. This chapter reviews the evidence supporting an association between pathology, large-scale chromatin organization, and nuclear architecture in this enigmatic syndrome.

Introduction

The ICF syndrome (OMIM 242860) is a rare genetic disorder with a distinctive chromosomal phenotype. The acronym ICF was coined more than 20 years ago (Maraschio et al. 1988) to describe a newly identified syndrome characterized

E.V. Volpi (✉)
Wellcome Trust Centre for Human Genetics, University of Oxford, Roosevelt Drive, Oxford OX3 7BN, UK
e-mail: emanuela.volpi@well.ox.ac.uk

Y.B. Yurov et al. (eds.), *Human Interphase Chromosomes: Biomedical Aspects*, DOI 10.1007/978-1-4614-6558-4_6, © Springer Science+Business Media, LLC 2013

by *i*mmunodeficiency, *c*entromeric instability, and *f*acial anomalies. Cytogenetically ICF is distinguishable because of its variable chromosomal instability and typical structural aberrations, mostly involving the centromeric regions of chromosomes 1 and 16, and to a lesser extent chromosome 9, with the most distinctive feature being the apparent stretching or despiralization of the large blocks of juxtacentromeric—or centromere-adjacent—heterochromatin. Chromosomal anomalies in ICF patients also include homologous and nonhomologous associations of those centromeric regions, multibranched configurations involving one or more of the decondensed chromosomes, nuclear protrusions, micronuclei, and duplications or deletions of whole chromosome arms (Brown et al. 1995; Gimelli et al. 1993; Sawyer et al. 1995a; Tuck-Muller et al. 2000). Those chromosomal abnormalities are almost exclusively encountered in phytohemagglutinin-stimulated lymphocytes. In fibroblasts and Epstein–Barr virus (EBV)-transformed lymphocytes, the cytogenetic manifestations of the syndrome are limited to the occasional elongation of the heterochromatic regions (Maraschio et al. 1989).

The ICF syndrome is often fatal in childhood. The immunodeficiency is typically the cause of early fatalities, and the normal cause of death in ICF patients is infection, usually of the pulmonary or gastrointestinal tract. Although the majority of patients display only humeral immunodeficiency, a considerable number show combined immunodeficiency with additional defective cellular immunity, characterized by a reduction or inversion in the $CD4^+/CD8^+$ T-cell ratio. Thus far, only one patient lacking the immunodeficiency phenotype has been described. Facial anomalies in ICF patients are mild and include hypertelorism, low-set ears, flat nasal bridge, epicanthal folds, and macroglossia. Psychomotor and mental retardation are also often observed in this syndrome. The spectrum of phenotypic and clinical features resulting in the pronounced heterogeneity of the ICF syndrome has been comprehensively covered in a number of expert reviews (Ehrlich 2003; Ehrlich et al. 2008; Hagleitner et al. 2008).

Since its initial description, the enigmatic nature of this syndrome has continued to fascinate scientists worldwide and from different biomedical spheres with the result that, in spite of its rarity—approximately 50 cases reported so far—and the limited availability of study material, published research on various aspects of ICF has flourished over the past 20 years. Although investigations on the ICF syndrome initially seemed almost exclusively the dominion of clinical genetics, cytogenetics, and immunology, with the growing number of case reports and the complexity of the disorder becoming gradually more evident, cross-referencing and the drawing of parallelisms between specific findings in ICF and related aspects of cancer biology, epigenetics, genomics, and developmental biology have become increasingly frequent (Aran et al. 2011; Ehrlich 2009; Ehrlich et al. 2006; Feng and Fan 2009; Martins-Taylor et al. 2012; Toyota and Suzuki 2010; van den Brand et al. 2011). However, notwithstanding the global research efforts and the extensive range of experimental approaches deployed, the full ICF pathogenetic picture remains somehow elusive.

Methylation Defects and Genetic Heterogeneity

Methylation at the 5-position of cytosine within CG dinucleotides is an important epigenetic modification, critical for gene regulation and control of chromatin structure in mammalian cells. One of the hallmark molecular features of the ICF syndrome is defective DNA methylation. Most specifically, ICF patients present with significant constitutive hypomethylation at the classical satellite 2 DNA, the major constituent of the juxtacentromeric heterochromatin at 1qh and 16qh, which is normally hypermethylated in somatic cells. Chromosome 9 juxtacentromeric heterochromatin, which mainly consists of the related satellite 3, also appears to be hypomethylated, although to a lesser extent (Jeanpierre et al. 1993). A small number of other genomic regions have been shown to have significant hypomethylation in ICF syndrome, most notably the centromeric alpha satellites (Miniou et al. 1997a), Alu sequences (Miniou et al. 1997b) and the non-satellite repeats D4Z4 and *NBL2* (Kondo et al. 2000). Single-copy loci showing heterogeneous hypomethylation comprise the imprinted loci D15S9, D15S63, and H19 (Schuffenhauer et al. 1995), and in female ICF cells a number of genes residing on the inactive X chromosome (Hansen et al. 2000; Bourc'his et al. 1999; Miniou et al. 1994).

Initial investigations on the DNA methylation status in ICF were prompted by the observation that chromosome rearrangements seen in ICF cells were reminiscent of the chromosomal changes induced by the undermethylating agent 5-azacytidine (examples in Fig. 6.1). Key cytological evidence implicating DNA hypomethylation in the formation of pericentromeric anomalies in ICF was provided by Ji et al. (1997), who confirmed in a pro-B-cell line the preferential formation at a very high frequency of pericentric rearrangements of chromosome 1—identical to the diagnostic chromosomal aberrations in ICF syndrome—on treatment with the DNA methylation inhibitors 5-azadeoxycytidine and 5-azacytidine.

At the molecular level, the methylation pattern of classical satellite DNA in ICF patients, which mimics an embryonic, undermethylated pattern, was initially shown

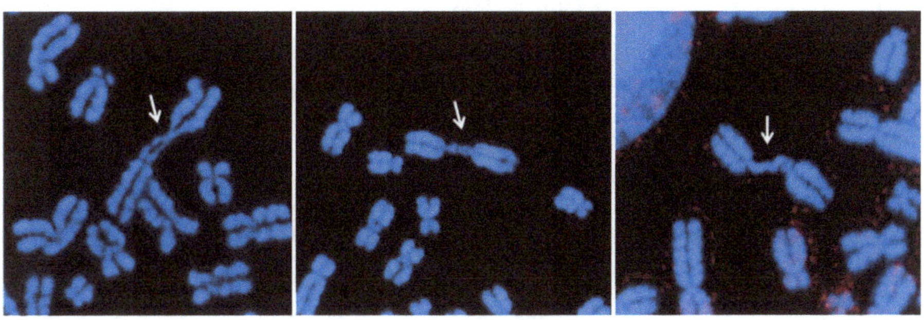

Fig. 6.1 Demethylation by 5-azacytidine induces ICF (immunodeficiency, centromere instability, and facial anomalies syndrome)-like 'stretching' of the juxtacentromeric heterochromatin in normal B cells

by restriction endonuclease analysis with methyl-sensitive enzymes (Jeanpierre et al. 1993). Subsequently, Hassan et al. (2001) managed to quantify the extent of the abnormal methylation in ICF by using a bisulfite conversion-based method and showing that the cytosine methylation level of the satellite 2 DNA sequences was almost 70 % in normal lymphoblasts and fibroblasts, compared with only 20 % in ICF cells. In line with previous findings, a paper recently published by the Esteller group (Heyn et al. 2012), reporting whole-genome bisulfite DNA sequencing of an ICF patient, showed that despite a global loss of DNA methylation, the shape of genetic features, such as promoters or CpG islands, is conserved, and the most profound changes occur in inactive heterochromatic regions, satellite repeats, and transposons.

The ICF syndrome—the transmission of which is compatible with an autosomal recessive mode of inheritance—was initially linked to chromosome 20q11.2 by homozygosity mapping (Wijmenga et al. 1998). Subsequently, the DNA methyltransferase 3B gene (*DNMT3B*) was mapped by positional cloning at this chromosomal location and mutations in this gene identified as responsible for the methylation defects observed in ICF (Hansen et al. 1999; Xu et al. 1999). Along with DNMT3A, DNMT3B acts to methylate cytosine residues de novo. The de novo methyltransferases DNMT3A and DNMT3B are essential for normal development, and *Dnmt3b*$^{-/-}$ homozygous mouse embryos present with multiple developmental defects and fail to develop to term, although development appears normal up to E9.5 (Okano et al. 1999).

Dnmt3b colocalizes with Dnmt3a to the pericentromeric heterochromatic regions in murine embryonic stem cells (Bachman et al. 2001). In mammalian cells, DNMT3B has been found to interact with constitutive centromere protein CENP-C to modulate DNA methylation and the histone code at centromeric regions, and its loss has been shown to lead to elevated chromosome misalignment and segregation defects during mitosis and increased transcription of centromeric repeats (Gopalakrishnan et al. 2009). DNMT3B has also been found to associate in heterochromatic regions with HDAC1 and HDAC2, the ATP-dependent chromatin remodeling enzyme hSNF2H, and two components of the histone H3 lysine 9 methylation machinery, namely, HP1 proteins and Suv39h1 (Geiman et al. 2004). In summary, DNMT3B appears to interact with four chromatin-associated enzymatic activities common to transcriptionally repressed, heterochromatic regions of the genome: DNA methyltransferase, histone deacetylase, ATPase, and histone methylase activities.

Mutations of *DNMT3B* in ICF syndrome are heterogeneous. Analysis of 14 patients revealed 11 different mutations, including eight different missense mutations, two nonsense mutations, and a splice site mutation (Wijmenga et al. 2000). Missense mutations of *DNMT3B* in ICF patients occur in or near the catalytic domain of the DNMT3B protein, presumably affecting the normal activity of the enzyme. Nonsense mutations giving rise to a truncated protein always occur as compound heterozygous mutations, highlighting that the DNMT3B protein is essential for life. Most recently, a murine model for ICF syndrome has been engineered by generating *Dnmt3b* mutations in mice (Ueda et al. 2006). Homozygous mice carrying two missense alleles of *Dnmt3b* show many ICF-like characteristics, and exhibit low body weight, cranial facial anomalies, and T-cell death by apoptosis. They also show hypomethylation of heterochromatin repeat DNA.

Mutations in the *DNMT3B* gene account for approximately 60 % of ICF cases. Wijmenga and collaborators were the first to report a number of ICF patients not carrying mutations in the *DNMT3B* gene (Wijmenga et al. 2000). Similar findings were subsequently published by other research groups (Jiang et al. 2005; Kubota et al. 2004). Jiang et al. (2005) showed that the patients carrying a mutation in the *DNMT3B* gene had alpha satellite methylation patterns comparable to control samples. In contrast, the patients who did not carry mutations in *DNMT3B* exhibited hypomethylation of the alpha satellite DNA as well as classical satellite. These findings led to the proposal of the existence of two distinct types of ICF syndrome with different genetic and epigenetic characteristics, namely, a type 1, in which patients display mutations in the *DNMT3B* gene but have normal alpha satellite methylation, and a type 2, characterized by normal *DNMT3B* and hypomethylation of alpha satellite DNA. There appears to be no genotype–phenotype correlation between patients with and without *DNMT3B* mutations (Hagleitner et al. 2008).

In 2011, by means of homozygosity mapping, whole-exome sequencing and Sanger sequencing on 11 ICF type 2 cases, de Greef and collaborators identified in some of the patients mutations in *ZBTB24*, the zinc-finger- and BTB (bric-a-bric, tramtrack, broad complex)-domain-containing 24, a gene belonging to a large family of transcriptional factors including some members with prominent regulatory roles in hematopoietic development and malignancy (de Greef et al. 2011). *ZBTB24* is ubiquitously expressed and its expression appears to be coregulated with *DNMT3B* during B-cell differentiation. A deletion in the *ZBTB24* gene, resulting in a loss of function of the corresponding protein, was also reported in a consanguineous Lebanese ICF type 2 family with three affected brothers (Chouery et al. 2012). In a *DNMT3B* mutant patient, whole-genome bisulfite DNA sequencing showed that *ZBTB24* harbored a hypermethylated DMR (differentially methylated region) in its promoter, suggesting that inactivating hypermethylation of this gene might contribute to the type 1 disease phenotype (Heyn et al. 2012).

Notwithstanding a great deal of progress with the understanding of the genetic and methylation defects that characterize the ICF syndrome at the molecular level, the causal relationships between *DNMT3B* and/or *ZBTB24* mutations, methylation abnormalities, and the range of ICF phenotypes still needs to be fully clarified.

Original Interphase Studies

Once it became clear that the ICF distinctive 'chromosomal phenotype' comprising anomalous mitotic configurations involving mainly chromosomes 1 and 16, and to a lesser extent chromosome 9, combined with an abnormal methylation pattern of the classical satellite DNA, was the hallmark feature of the ICF syndrome, there followed several attempts at investigating the ICF genetic and cytogenetic traits in interphase as well as in metaphase. Of course, by that time, the introduction of the fluorescence in situ hybridization (FISH) technique to the diagnostic scenario had allowed the reinvention of classical cytogenetics into molecular cytogenetics, opening up the possibility to visualize specific DNA sequences within the nuclear context and allowing the identification of individual chromosomes and the study chromosomal abnormalities in interphase (Volpi and Bridger 2008).

Maraschio and collaborators published the first study on the interphase behavior of the centromeric heterochromatin of chromosomes 1 and 16 in ICF lymphocytes by means of nonisotopic in situ hybridization, reporting involvement of those chromosomal regions in increased somatic pairing, nuclear protrusions, and micronuclei (Maraschio et al. 1992). Miniou and collaborators carried out the first cytological investigation of DNA methylation defects in ICF using 5-methylcytosine monoclonal antibody to evaluate the methyl content of nuclei and micronuclei (Miniou et al. 1994). They also used alpha and classical satellite probes in single- and dual-color FISH to visualize heterochromatin stretching in interphase. Sawyer and collaborators applied FISH with 'painting' probes for chromosomes 1 and 16 to document the progression of centromeric instability from simple heterochromatin decondensation to the subsequent formation of multibranched configurations, and finally nuclear projections and micronuclei involving both chromosomes 1 and 16 (Sawyer et al. 1995b). Similar interphase cytogenetic investigations were carried out by Stacey et al. (1995). Subsequent FISH experiments using the satellite 2-related probe pHuR 195 confirmed that chromosome fusion in the ICF syndrome occurs only at regions of decondensed centromere-adjacent heterochromatin, and that the alpha satellite repeats, the main component of centromeres, always remain outside the regions of multiradiate chromosome fusions (Sumner et al. 1998).

Gisselsson and collaborators showed, in four ICF lymphoblastoid cell lines, an increased colocalization of the hypomethylated 1qh and 16qh sequences in interphase, abnormal looping of pericentromeric DNA sequences at metaphase, formation of bridges at anaphase, chromosome 1 and 16 fragmentation at the telophase–interphase transition, and, in apoptotic cells, micronuclei with overrepresentation of chromosome 1 and 16 material. Their results suggested that 1qh–16qh associations in interphase can lead to disturbances of mitotic segregation, resulting in micronucleus formation and sometimes apoptosis (Gisselsson et al. 2005). An association between satellite 2 demethylation induced by 5-azacytidine with missegregation of chromosomes 1 and 16 was shown in human somatic cells by Prada et al. (2012).

Overall, the original interphase cytogenetic studies were mostly concerned with tracking the succession of cytological events leading to the distinctive karyotypic features of the ICF syndrome. Accordingly, the main focus of those research endeavors was on the dynamics of chromosomal instability and references to the possible pathogenetic significance of interphase chromosome organization were rare and strictly speculative. However, those initial 'nuclear' observations were fundamental to prepare the experimental and conceptual ground for subsequent studies on the epigenetic impact of nuclear architecture and interphase chromosome organization in the ICF syndrome.

Gene Expression and the Regulation Conundrum

A significant landmark in the sequence of research undertakings aimed at clarifying the connection between methylation defects and gene expression in ICF was a paper published by Ehrlich and collaborators in 2001 (Ehrlich et al. 2001). By means of an oligonucleotide microarray containing approximately 5,600 different human genes, gene expression was analyzed in lymphoblastoid cell lines from six ICF patients carrying different *DNMT3B* mutations. A total of 45 genes were shown to have significant differences in expression in ICF versus control cells. Of the 32 genes described in detail in the paper, a substantial number appears to play a part in lymphogenesis, signal transduction, and apoptosis. Nine genes in particular are implicated in lymphoid cell differentiation or function after V(D)J recombination and are involved in signal transduction or transcriptional control. Despite the differences in gene expression between ICF and control lymphoblastoid cell lines, no differences in methylation at the promoters or 5 -transcribed regions were observed in a sample of genes analyzed by COBRA analysis, leading the authors to suggest that the hypomethylation of classical satellite DNA sequences in ICF syndrome might result in a decreased 'heterochromatinization' that affects the regulation of genes elsewhere in the genome.

Global expression profiling of three patients with *DNMT3B* mutations was published by Jin et al. (2008). This study generated an impressive list of nearly 800 genes appearing to be up- or downregulated in ICF cells, a large number of which are involved with immune function, development, and neurogenesis. Detailed DNA methylation mapping on a subset of deregulated genes revealed that a fraction of them are typically methylated at low levels in normal cells and lose their methylation in ICF cells. By chromatin immune precipitation the authors also showed that histone modification patterns at affected promoters were dramatically altered, demonstrating for the first time that loss of DNMT3B function in ICF can lead to significant hypomethylation of nonrepetitive regions of the genome and alterations in the histone code. They also observed loss of binding of the SUZ12 component of the PRC2 polycomb repression complex and DNMT3B to derepressed genes, including a number of homeobox genes critical to the immune system and craniofacial development.

Heterochromatic genes have also been reported to undergo epigenetic changes and escape silencing in ICF (Brun et al. 2011); however, the precise mechanism driving their transcription is unclear. Indeed, although DNA hypomethylation was found in all heterochromatic genes and in all patients analyzed, gene expression was restricted to some genes and every patient had his or her own group of activated genes. Surprisingly, heterochromatic genes—either active or inactive—appeared to be associated with histone modifications typical of inactive chromatin. Hypomethylation of subtelomeric regions, associated with aberrant transcription and advanced replication timing of these sequences, has also been described in ICF (Yehezkel et al. 2008).

All together, these expression studies have succeeded in shedding important light on the extent and nature of gene deregulation in ICF resulting from the combined effect of DNA hypomethylation and alterations in the histone code. However, even allowing for prediction of alterations in relevant downstream targets of misexpressed transcriptional regulators, it is obvious that the full ICF epigenetic picture is not yet complete. In summary, on the basis of what has been published so far, and taking also in consideration recent evidence obtained by whole-genome bisulfite DNA sequencing of a DNMT3B mutant patient showing that, despite global loss of genomic DNA methylation, transcriptional active loci and rRNA repeats escape hypomethylation and overall the shape of genetic features such as promoters and CpG islands is conserved (Heyn et al. 2012), it is apparent that the relationship between methylation defects, altered gene expression, and clinical and phenotypic features in ICF needs further investigation.

The Nuclear Architecture Perspective

Over the years a number of hypotheses advocating an epigenetic model for the ICF syndrome have been advanced by different research groups. In particular, possible mechanisms linking altered gene expression to the hypomethylation of pericentromeric heterochromatin have been postulated, generally envisaging inappropriate release or recruitment of regulatory complexes by the hypomethylated satellite DNA, affecting the transacting regulatory properties of the heterochromatin (Hassan et al. 2001; Hansen et al. 1999; Xu et al. 1999; Ehrlich et al. 2001; Bickmore and van der Maarel 2003).

Although so far only a handful of investigations have attempted to test those hypotheses, some interesting findings have already emerged. For example, two publications have independently reported repositioning of the chromosome 1 juxtacentromeric heterochromatin domain to a more interior location within the nuclear volume in ICF lymphocytes (Dupont et al. 2012; Jefferson et al. 2010). More specifically, microscope observations on a cell-by-cell basis, and comparative measurements of the distance between chromosome 1 juxtacentromeric heterochromatin and the nuclear rim carried out in our laboratory, have revealed that the extent of heterochromatin association with the extreme nuclear periphery is reduced in ICF cells (both type 1 and type 2), suggesting its specific repositioning to a more internal location within the nuclear space (Jefferson et al. 2010). Dupont and collaborators have independently confirmed repositioning of the chromosome 1 juxtacentromeric heterochromatin toward the nuclear interior by means of three-dimensional (3D) FISH and confocal microscopy analysis on type 2 patient cells (Dupont et al. 2012). Based on the differential distribution of early and late-replicating chromatin within the nuclear space (Ferreira et al. 1997; Sadoni et al. 1999), these findings on the relocation of the juxtacentromeric heterochromatin away from the extreme nuclear periphery to a more internal position appear to be in line with the previously reported data on the advanced replication of the hypomethylated satellite 2 in ICF (Hassan et al. 2001).

Although its role in actively regulating gene expression remains unproven, the nuclear periphery is generally considered a transcriptionally silent ‘address’ within the nuclear volume, characterized by poor gene density (Croft et al. 1999; Shopland et al. 2006; Tanabe et al. 2002) and a high concentration of nontranscribed sequences (Scheuermann et al. 2004). Interestingly, a case of functional repositioning of the chromosome 1 juxtacentromeric heterochromatin within the nuclear volume had been reported previously by Barki-Celli et al. (2005), who found that the 1q12 heterochromatin domain location within the nuclear volume was different when resting B cells were compared to B cells treated with trichostatin A (TSA), a histone deacetylase inhibitor, with the induced hyperacetylation shown to cause a significant reduction in the percentage of 1q12 signals associated with the nuclear periphery, probably by interfering with the establishment of 'epigenetic codes' required for localizing 1q12 to the nuclear membrane, as suggested by the authors.

Long-range heterochromatin–gene associations have gained recognition as a potential novel epigenetic mechanism of gene regulation in mammalian cells. So far, a correlation between gene silencing and localization to transcriptionally repressive heterochromatic compartments has been reported in mouse cycling lymphocytes (Brown et al. 1997, 1999; Grogan et al. 2001), human and mouse erythroid cells (Francastel et al. 1999, 2001; Schubeler et al. 2000), and retinoblastoma cells (Bartova et al. 2002). Replacement of the endogenous β-globin locus control region (LCR) with an ectopic regulatory element, the IgH LCR, was shown to induce looping of the globin locus out of its chromosome territory and bring it closer to the repressive centromeric heterochromatin compartment by Ragoczy et al. (2003). A link between centromeric recruitment and establishment of allelic exclusion at the immunoglobulin heavy-chain gene in mouse B cells was also reported (Roldan et al. 2005). In many instances, uncertainty remains whether nonrandom positioning and association of genes with specific nuclear compartments are the cause or the consequence of gene function, although the general consensus points toward an interactive model in which the functional potential of a locus facilitates its association with a nuclear compartment, which in turn influences the functional properties of the locus (Misteli 2004). Accordingly, recruitment to heterochromatin domains may help to stabilize the inactive state, rather than actively promoting silencing.

In our laboratory, we were interested in understanding if the decondensation of the juxtacentromeric heterochromatin, as observed in metaphase, and general chromosomal instability reported in ICF patients, corresponded to changes in the three-dimensional properties of the heterochromatin in interphase, our working hypothesis being that disruption in the heterochromatin spatial configuration might interfere with transcriptional silencing and be indirectly responsible for the changes in gene expression accounting for the symptoms of ICF. Accordingly, as well as the heterochromatin intranuclear positioning, we analyzed and compared, in patients and controls, the large-scale organization of the juxtacentromeric heterochromatin of chromosome 1 and the intranuclear positioning of a set of genes—four from chromosome 1 and one as a control from chromosome 6, on which we had concurrently conducted expression and methylation analysis—and their colocalization with the juxtacentromeric heterochromatin of chromosome 1. Examples of our findings are presented in Figs. 6.2 and 6.3. In contrast to the consensual view that decondensation and stretching of the juxtacentromeric heterochromatic blocks—similar to that

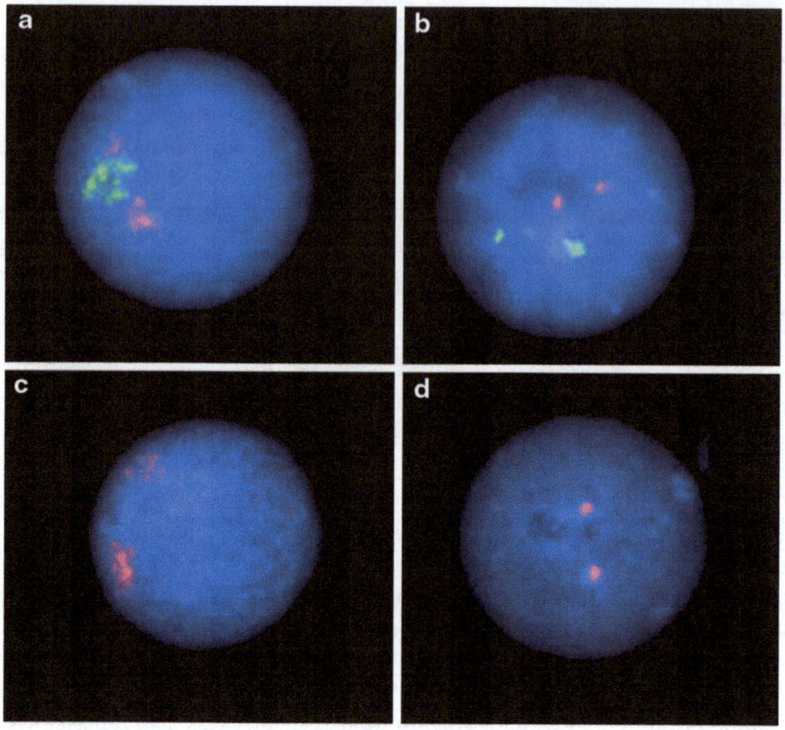

Fig. 6.2 The intranuclear positioning and large-scale organization of the juxtacentromeric heterochromatin is altered in ICF B cells. Chromosome 1 heterochromatin is visualized by two-dimensional (2D) fluorescence in situ hybridization (FISH) with the classical satellite probes D1Z1 (*red*) in nuclei from two controls (**a**, **c**), an ICF type 1 patient (**b**), and an ICF type 2 patient (**d**). Chromosome 16 heterochromatin is visualized by cohybridization with D16Z3 (*green*) (**a**, **b**)

observed in ICF cells in metaphase—could also be expected in interphase, we found that in ICF nuclei the juxtacentromeric heterochromatin appears more compactly shaped. We also found that for two of the loci analyzed—*CNN3* and *RGS1*—the extent of gene–heterochromatin colocalization was significantly reduced in ICF cells, in which these genes are overexpressed but present no changes in promoter methylation (Jefferson et al. 2010).

Our combined findings on the altered large-scale organization and intranuclear positioning of chromosome 1 juxtacentromeric heterochromatin in ICF are particularly significant in the light of the mounting experimental evidence suggesting chromosome band 1q12, which corresponds to the juxtacentromeric heterochromatin of chromosome 1, to be the core of a nuclear domain with functional significance. Earlier investigations showed physical association of this genomic region with the human polycomb group complex, a repressor of the homeotic gene expression (Saurin et al. 1998), and also with the oncogenic transcriptional regulator TLX1/HOX11 in leukemic T cells (Heidari et al. 2006). Most relevantly, in ICF cells the

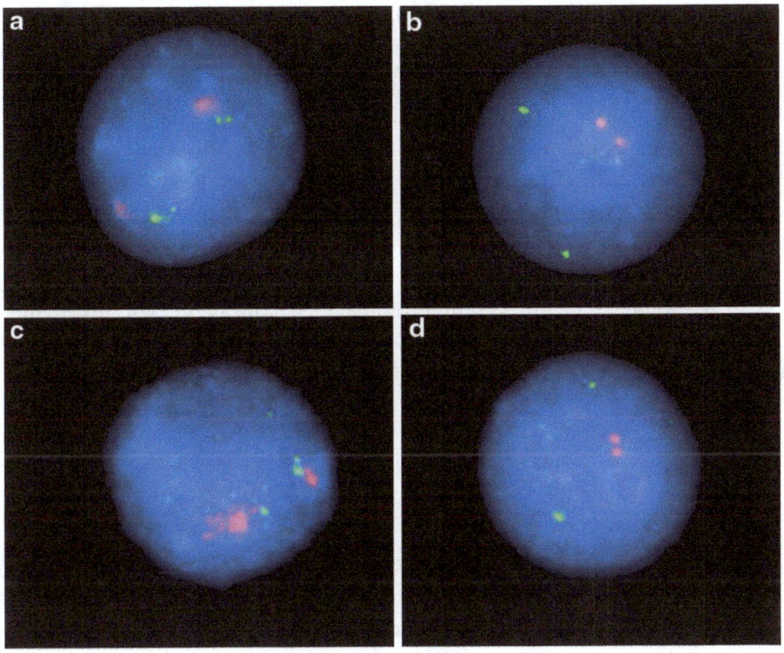

Fig. 6.3 The extent of interphase gene–heterochromatin colocalization for the two loci RGS1 (*green*) (**a**, **b**) and CNN3 (*green*) (**c**, **d**) is significantly reduced in ICF cells (**b**, **d**), in which they are significantly overexpressed, when compared to respective controls (**a**, **c**). Chromosome 1 heterochromatin is visualized in *red* by hybridization with D1Z1 in all four images

subcellular distribution of HP1 proteins is altered and the 1qh satellite DNA is associated in G_2 with a giant HP1-PML nuclear body (Luciani et al. 2005, 2006).

A large-scale effect of DNA hypomethylation on the organization of chromosome 'territories' within the nuclear volume in ICF was proposed by Matarazzo et al. (2007) on the basis of their findings on repositioning of misexpressed genes located in the pseudo-autosomal region 2 (PAR2) of the X and Y chromosomes. As DNA hypomethylation appears to cause changes in the interphase organization of the inactive chromosome X territory that extends far beyond the genes which are immediately subject to hypomethylation and transcriptional activation, they suggest that it is plausible that elsewhere in the ICF genome, altered chromosome territory organization encompassing extended regions around hypomethylated sequences may allow for the inappropriate transcriptional activation of relocated genes if the right transcription factor environment is available.

Finally, DNA replication appears to be globally affected in ICF cells. Gartler and collaborators (Hansen et al. 2000) indicated advanced replication time as a major determinant of escape from silencing for genes on the inactive X and Y chromosomes they had identified in ICF syndrome. They also showed that satellite 2 replication in ICF is advanced compared with that in normal cells (Hassan et al. 2001). Yehezkel and collaborators reported that hypomethylation of subtelomeric regions was associated in ICF with advanced telomere replication timing (Yehezkel et al. 2008). More recently, the De Sario group reported that in ICF cells heterochromatic genes replicate earlier in the S-phase, global replication fork speed is higher, and S-phase is shorter, and suggested that these replication defects may result from chromatin changes which modify DNA accessibility to the replication machinery (Lana et al. 2012).

From what has been published so far on different aspects of interphase organization in ICF cells, a distinctive ICF 'nuclear phenotype' has begun to emerge in which both local and global changes in the chromatin large-scale organization and positioning, together with changes in chromatin proteins distribution, appear mechanistically linked to the deregulation or impairment of fundamental nuclear functions, including chromosome condensation and segregation, DNA replication, and transcription.

Conclusion

The complexity of ICF, in particular the combination of phenotypic variability and genetic heterogeneity that characterizes it, has intrigued geneticists and cell biologists since this syndrome was initially described more than 20 years ago. Indeed, irrespective of its rare occurrence, ICF has been the object of numerous and diverse studies that have yielded interesting insights into its pathogenesis and prompted a substantial amount of speculation on the relationship between methylation defects and the ICF phenotypic spectrum. Up to now only a small number of investigations have attempted to study the ICF syndrome from a nuclear architecture perspective. However, findings that have emerged so far appear to support—cautiously but consistently—a role for large-scale chromatin organization in the interphase nucleus as a structural and functional intermediary between DNA hypomethylation and altered gene expression in ICF, suggesting the existence of additional layers of epigenetic complexity in this syndrome and warranting further investigations.

Acknowledgments Dr. E.V. Volpi thanks the Wellcome Trust for support [075491/Z/04].

Some of the images used in Figs. 6.1, 6.2, and 6.3 were reproduced from Jefferson et al. (2010), *PLoS ONE* 5(6):e11364, an open-access article distributed under the terms of the Creative Commons Attribution License.

References

Aran D et al (2011) Replication timing-related and gene body-specific methylation of active human genes. Hum Mol Genet 20(4):670–680

Bachman KE, Rountree MR, Baylin SB (2001) Dnmt3a and Dnmt3b are transcriptional repressors that exhibit unique localization properties to heterochromatin. J Biol Chem 276(34):32282–32287

Barki-Celli L et al (2005) Differences in nuclear positioning of 1q12 pericentric heterochromatin in normal and tumor B lymphocytes with 1q rearrangements. Genes Chromosomes Cancer 43(4):339–349

Bartova E et al (2002) Nuclear structure and gene activity in human differentiated cells. J Struct Biol 139(2):76–89
Bickmore WA, van der Maarel SM (2003) Perturbations of chromatin structure in human genetic disease: recent advances. Hum Mol Genet 2:5
Bourc'his D et al (1999) Abnormal methylation does not prevent X inactivation in ICF patients. Cytogenet Cell Genet 84(3–4):245–252
Brown DC et al (1995) ICF syndrome (immunodeficiency, centromeric instability and facial anomalies): investigation of heterochromatin abnormalities and review of clinical outcome. Hum Genet 96(4):411–416
Brown KE et al (1997) Association of transcriptionally silent genes with Ikaros complexes at centromeric heterochromatin. Cell 91(6):845–854
Brown KE et al (1999) Dynamic repositioning of genes in the nucleus of lymphocytes preparing for cell division. Mol Cell 3(2):207–217
Brun ME et al (2011) Heterochromatic genes undergo epigenetic changes and escape silencing in immunodeficiency, centromeric instability, facial anomalies (ICF) syndrome. PLoS One 6(4):e19464
Chouery E et al (2012) A novel deletion in ZBTB24 in a Lebanese family with immunodeficiency, centromeric instability, and facial anomalies syndrome type 2. Clin Genet 82(5):489–493
Croft JA et al (1999) Differences in the localization and morphology of chromosomes in the human nucleus. J Cell Biol 145(6):1119–1131
de Greef JC et al (2011) Mutations in ZBTB24 are associated with immunodeficiency, centromeric instability, and facial anomalies syndrome type 2. Am J Hum Genet 88(6):796–804
Dupont C et al (2012) 3D position of pericentromeric heterochromatin within the nucleus of a patient with ICF syndrome. Clin Genet 82(2):187–192
Ehrlich M (2003) The ICF syndrome, a DNA methyltransferase 3B deficiency and immunodeficiency disease. Clin Immunol 109(1):17–28
Ehrlich M (2009) DNA hypomethylation in cancer cells. Epigenomics 1(2):239–259
Ehrlich M et al (2001) DNA methyltransferase 3B mutations linked to the ICF syndrome cause dysregulation of lymphogenesis genes. Hum Mol Genet 10(25):2917–2931
Ehrlich M, Jackson K, Weemaes C (2006) Immunodeficiency, centromeric region instability, facial anomalies syndrome (ICF). Orphanet J Rare Dis 1:2
Ehrlich M et al (2008) ICF, an immunodeficiency syndrome: DNA methyltransferase 3B involvement, chromosome anomalies, and gene dysregulation. Autoimmunity 41(4):253–271
Feng J, Fan G (2009) The role of DNA methylation in the central nervous system and neuropsychiatric disorders. Int Rev Neurobiol 89:67–84
Ferreira J et al (1997) Spatial organization of large-scale chromatin domains in the nucleus: a magnified view of single chromosome territories. J Cell Biol 139(7):1597–1610
Francastel C et al (1999) A functional enhancer suppresses silencing of a transgene and prevents its localization close to centromeric heterochromatin. Cell 99(3):259–269
Francastel C, Magis W, Groudine M (2001) Nuclear relocation of a transactivator subunit precedes target gene activation. Proc Natl Acad Sci U S A 98(21):12120–12125
Geiman TM et al (2004) DNMT3B interacts with hSNF2H chromatin remodeling enzyme, HDACs 1 and 2, and components of the histone methylation system. Biochem Biophys Res Commun 318(2):544–555
Gimelli G et al (1993) ICF syndrome with variable expression in sibs. J Med Genet 30(5):429–432
Gisselsson D et al (2005) Interphase chromosomal abnormalities and mitotic missegregation of hypomethylated sequences in ICF syndrome cells. Chromosoma (Berl) 114(2):118–126
Gopalakrishnan S et al (2009) DNMT3B interacts with constitutive centromere protein CENP-C to modulate DNA methylation and the histone code at centromeric regions. Hum Mol Genet 18(17):3178–3193
Grogan JL et al (2001) Early transcription and silencing of cytokine genes underlie polarization of T helper cell subsets. Immunity 14(3):205–215

Hagleitner MM et al (2008) Clinical spectrum of immunodeficiency, centromeric instability and facial dysmorphism (ICF syndrome). J Med Genet 45(2):93–99

Hansen RS et al (1999) The DNMT3B DNA methyltransferase gene is mutated in the ICF immunodeficiency syndrome. Proc Natl Acad Sci U S A 96(25):14412–14417

Hansen RS et al (2000) Escape from gene silencing in ICF syndrome: evidence for advanced replication time as a major determinant. Hum Mol Genet 9(18):2575–2587

Hassan KM et al (2001) Satellite 2 methylation patterns in normal and ICF syndrome cells and association of hypomethylation with advanced replication. Hum Genet 109(4):452–462

Heidari M et al (2006) The nuclear oncoprotein TLX1/HOX11 associates with pericentromeric satellite 2 DNA in leukemic T-cells. Leukemia 20(2):304–312

Heyn H et al (2012) Whole-genome bisulfite DNA sequencing of a DNMT3B mutant patient. Epigenetics 7:6

Jeanpierre M et al (1993) An embryonic-like methylation pattern of classical satellite DNA is observed in ICF syndrome. Hum Mol Genet 2(6):731–735

Jefferson A et al (2010) Altered intra-nuclear organisation of heterochromatin and genes in ICF syndrome. PLoS One 5(6):e11364

Ji W et al (1997) DNA demethylation and pericentromeric rearrangements of chromosome 1. Mutat Res 379(1):33–41

Jiang YL et al (2005) DNMT3B mutations and DNA methylation defect define two types of ICF syndrome. Hum Mutat 25(1):56–63

Jin B et al (2008) DNA methyltransferase 3B (DNMT3B) mutations in ICF syndrome lead to altered epigenetic modifications and aberrant expression of genes regulating development, neurogenesis and immune function. Hum Mol Genet 17(5):690–709

Kondo T et al (2000) Whole-genome methylation scan in ICF syndrome: hypomethylation of non-satellite DNA repeats D4Z4 and NBL2. Hum Mol Genet 9(4):597–604

Kubota T et al (2004) ICF syndrome in a girl with DNA hypomethylation but without detectable DNMT3B mutation. Am J Med Genet A 1(3):290–293

Lana E et al (2012) DNA replication is altered in immunodeficiency centromeric instability facial anomalies (ICF) cells carrying DNMT3B mutations. Eur J Hum Genet 29(10):41

Luciani JJ et al (2005) Subcellular distribution of HP1 proteins is altered in ICF syndrome. Eur J Hum Genet 13(1):41–51

Luciani JJ et al (2006) PML nuclear bodies are highly organised DNA-protein structures with a function in heterochromatin remodelling at the G2 phase. J Cell Sci 119(pt 12):2518–2531

Maraschio P et al (1988) Immunodeficiency, centromeric heterochromatin instability of chromosomes 1, 9, and 16, and facial anomalies: the ICF syndrome. J Med Genet 25(3):173–180

Maraschio P et al (1989) Differential expression of the ICF (immunodeficiency, centromeric heterochromatin, facial anomalies) mutation in lymphocytes and fibroblasts. J Med Genet 26(7):452–456

Maraschio P et al (1992) Interphase cytogenetics of the ICF syndrome. Ann Hum Genet 56(pt 3):273–278

Martins-Taylor K et al (2012) Role of DNMT3B in the regulation of early neural and neural crest specifiers. Epigenetics 7:1

Matarazzo MR et al (2007) Chromosome territory reorganization in a human disease with altered DNA methylation. Proc Natl Acad Sci U S A 104(42):16546–16551

Miniou P et al (1994) Abnormal methylation pattern in constitutive and facultative (X inactive chromosome) heterochromatin of ICF patients. Hum Mol Genet 3(12):2093–2102

Miniou P et al (1997a) Alpha-satellite DNA methylation in normal individuals and in ICF patients: heterogeneous methylation of constitutive heterochromatin in adult and fetal tissues. Hum Genet 99(6):738–745

Miniou P et al (1997b) Undermethylation of Alu sequences in ICF syndrome: molecular and in situ analysis. Cytogenet Cell Genet 77(3–4):308–313

Misteli T (2004) Spatial positioning; a new dimension in genome function. Cell 119(2):153–156

Okano M et al (1999) DNA methyltransferases Dnmt3a and Dnmt3b are essential for de novo methylation and mammalian development. Cell 99(3):247–257
Prada D et al (2012) Satellite 2 demethylation induced by 5-azacytidine is associated with missegregation of chromosomes 1 and 16 in human somatic cells. Mutat Res 729(1–2):100–105
Ragoczy T et al (2003) A genetic analysis of chromosome territory looping: diverse roles for distal regulatory elements. Chromosome Res 11(5):513–525
Roldan E et al (2005) Locus 'decontraction' and centromeric recruitment contribute to allelic exclusion of the immunoglobulin heavy-chain gene. Nat Immunol 6(1):31–41
Sadoni N et al (1999) Nuclear organization of mammalian genomes. Polar chromosome territories build up functionally distinct higher order compartments. J Cell Biol 146(6):1211–1226
Saurin AJ et al (1998) The human polycomb group complex associates with pericentromeric heterochromatin to form a novel nuclear domain. J Cell Biol 142(4):887–898
Sawyer JR et al (1995a) Chromosome instability in ICF syndrome: formation of micronuclei from multibranched chromosomes 1 demonstrated by fluorescence in situ hybridization. Am J Med Genet 56(2):203–209
Sawyer JR et al (1995b) Centromeric instability of chromosome 1 resulting in multibranched chromosomes, telomeric fusions, and "jumping translocations" of 1q in a human immunodeficiency virus-related non-Hodgkin's lymphoma. Cancer (Phila) 76(7):1238–1244
Scheuermann MO et al (2004) Topology of genes and nontranscribed sequences in human interphase nuclei. Exp Cell Res 301(2):266–279
Schubeler D et al (2000) Nuclear localization and histone acetylation: a pathway for chromatin opening and transcriptional activation of the human beta-globin loc.s. Genes Dev 14(8):940–950
Schuffenhauer S et al (1995) DNA, FISH and complementation studies in ICF syndrome: DNA hypomethylation of repetitive and single copy loci and evidence for a trans acting factor. Hum Genet 96(5):562–571
Shopland LS et al (2006) Folding and organization of a contiguous chromosome region according to the gene distribution pattern in primary genomic sequence. J Cell Biol 174(1):27–38
Stacey M, Bennett MS, Hulten M (1995) FISH analysis on spontaneously arising micronuclei in the ICF syndrome. J Med Genet 32(7):502–508
Sumner AT, Mitchell AR, Ellis PM (1998) A FISH study of chromosome fusion in the ICF syndrome: involvement of paracentric heterochromatin but not of the centromeres themselves. J Med Genet 35(10):833–835
Tanabe H et al (2002) Non-random radial arrangements of interphase chromosome territories: evolutionary considerations and functional implications. Mutat Res 504(1–2):37–45
Toyota M, Suzuki H (2010) Epigenetic drivers of genetic alterations. Adv Genet 70:309–323
Tuck-Muller CM et al (2000) DNA hypomethylation and unusual chromosome instability in cell lines from ICF syndrome patients. Cytogenet Cell Genet 89(1–2):121–128
Ueda Y et al (2006) Roles for Dnmt3b in mammalian development: a mouse model for the ICF syndrome. Development (Camb) 133(6):1183–1192
van den Brand M et al (2011) Angiosarcoma in a patient with immunodeficiency, centromeric region instability, facial anomalies (ICF) syndrome. Am J Med Genet A 3:622–625
Volpi EV, Bridger JM (2008) FISH glossary: an overview of the fluorescence in situ hybridization technique. Biotechniques 45(4):385–386, 388, 390, passim
Wijmenga C et al (1998) Localization of the ICF syndrome to chromosome 20 by homozygosity mapping. Am J Hum Genet 63(3):803–809
Wijmenga C et al (2000) Genetic variation in ICF syndrome: evidence for genetic heterogeneity. Hum Mutat 16(6):509–517
Xu GL et al (1999) Chromosome instability and immunodeficiency syndrome caused by mutations in a DNA methyltransferase gene. Nature (Lond) 402(6758):187–191
Yehezkel S et al (2008) Hypomethylation of subtelomeric regions in ICF syndrome is associated with abnormally short telomeres and enhanced transcription from telomeric regions. Hum Mol Genet 17(18):2776–2789

Chapter 7
Interphase Cytogenetics at the Earliest Stages of Human Development

Darren K. Griffin, Gothami Fonseka, Helen G. Tempest, Alan R. Thornhill, and Dimitris Ioannou

Introduction

The widespread use of in vitro fertilization (IVF) throughout the world provides the opportunity to study human development at the very earliest stages before implantation. Nonetheless, the study of human embryos poses a series of unique ethical and moral implications. The unique totipotent nature of a human embryo and its potential to develop into a child necessitates a level of restriction and regulatory control that is not present when studying other cell types. Although some governments outlaw any experimental procedure on human embryonic material, others allow it under appropriate control. In the latter case (e.g., in the UK), experimentation can be justified on the basis of development of a diagnostic test and/or the goal of improving patient care. A further challenge to effective study is the paucity of material available. Much of the work reported in this chapter arises from the study of only single nuclei. For these reasons, research on interphase cytogenetics in human preimplantation embryos is less advanced than in other cell types. Despite this, a fundamental insight into chromosome copy number and nuclear organization can be gleaned from this material through collaboration with an appropriate clinical program. As attested by other chapters in this book, fluorescence in situ hybridization (FISH) was first adopted for research, but clinical applications rapidly followed. Prenatal and cancer diagnostics are the best examples of this, but the

D.K. Griffin (✉) • G. Fonseka
School of Biosciences, University of Kent, Canterbury, UK
e-mail: d.k.griffin@kent.ac.uk

H.G. Tempest • D. Ioannou
Department of Human and Molecular Genetics, Florida International University, Miami, FL, USA

A.R. Thornhill
The London Bridge Fertility, Gynaecology and Genetics Centre, London, UK

Y.B. Yurov et al. (eds.), *Human Interphase Chromosomes: Biomedical Aspects*, DOI 10.1007/978-1-4614-6558-4_7, © Springer Science+Business Media, LLC 2013

increasing use of assisted reproductive technologies, namely IVF, precipitated the use of FISH in the field of preimplantation genetic diagnosis (PGD). PGD is defined as the diagnosis of genetic disorders in human preimplantation embryos. The purpose is selective implantation of unaffected embryos in the hope of establishing genetically normal ongoing pregnancies. PGD by interphase cytogenetics was first applied for sexing (to screen for sex-linked disorders), then for chromosome translocations, and later for chromosome copy number. In the latter case, termed preimplantation genetic screening (PGS), families at risk of adverse obstetrical outcomes (referral categories include advanced maternal age and recurrent miscarriage) are targeted, rather than families at risk of transmitting inherited disorders in a classical Mendelian fashion. Clinical application of interphase cytogenetics in the IVF world has allowed the subsequent study of chromosome copy number and nuclear organization. This chapter provides an overview of interphase cytogenetics in human embryos, highlighting the progress and the sometimes contentious pitfalls that it has encountered.

Early Detection of the Sex Chromosomes

The first application of FISH in IVF for PGD was as a means of determining the genotypic sex of embryos from couples in whom there was a high risk of transmitting X-linked disorders. Application of dual-color FISH probes for the sex chromosomes was applied to interphase nuclei of single blastomeres (Griffin et al. 1991, 1992), which resulted in the first successful clinical PGD application and led to the birth of a healthy female child (Griffin et al. 1993). The advantage of the technique was the simultaneous detection of the sex chromosomes and the additional information about the copy number, which was not possible via polymerase chain reaction (PCR) at the time (Griffin 1994). PGD by FISH was rapidly adopted by IVF clinics worldwide for sex-linked disorders, and the possibility to select chromosomally normal embryos to improve IVF and ICSI pregnancies was proposed. This technology is still in use today, but in the past few years is being replaced with newer platforms that are discussed later in this chapter.

Rapidly technological advances quickly revolutionized the field, including (1) development of probes directly labeled with fluorochromes, (2) discovery and application of multiple fluorochromes enabling multicolor FISH experiments (up to six colors applied per experiment, and the cells can be reprobed (Ioannou et al. 2011a), (3) availability of commercial FISH probes, and (4) improvements in embryo fixation techniques. All these advances allowed shorter protocol times (Harper et al. 1994), improved diagnostic test results, and enabled multiple chromosomes other than just the sex chromosomes to be tested, thus facilitating the screening of chromosome translocations and aneuploidy.

Detection of Translocations in Early Human Preimplantation Embryos

FISH was soon applied in the field of PGD for the detection of the unbalanced products of chromosome translocations. The frequency of a balanced translocation in the general population is estimated to occur in 1 in 625 individuals, whereas in couples with recurrent miscarriage the incidence is much higher, affecting approximately 4–5 %, including both reciprocal and Robertsonian translocations (Chang et al. 2011). A balanced translocation is rarely an issue for the individual carrying it; however, problems arise during meiosis when the homologous chromosomes must pair, leading to the formation of either a quadrivalent or trivalent and subsequent aberrant segregation patterns in gametes that can be detected by interphase cytogenetics.

Scriven et al. (1998) estimated that for reciprocal translocations the chance for normal or balanced gametes is 4 in 32 whereas it is 4 in 16 for Robertsonian translocations. The chromosomes involved in the translocation, the translocation breakpoints, and the sex of the carrier are additional factors that can affect the percentage of the normal or balanced gametes and the risk of liveborn affected offspring. Carriers of a balanced translocation are therefore at an increased risk for infertility, pregnancy loss, or offspring with congenital abnormalities and mental retardation caused by unbalanced gamete segregation (Chang et al. 2011). PGD through interphase cytogenetics has provided a means to exclude embryos carrying an unbalanced chromosome complement.

Initially, FISH using whole chromosome paints was used to determine the origin of the additional chromosomal segment to clarify the karyotype and provide clinically important information (Mewar et al. 1992); however, this strategy was limited in its resolution. Another strategy used probes that spanned or flanked the chromosome breakpoints involved in the translocation to distinguish normal from balanced or unbalanced embryos, but this methodology was restricted by the cost and the time required to make specific probes (Chang et al. 2011). A more efficient approach involved the use of subtelomeric probes in combination with centromeric probes for specific chromosome pair combinations that does not require the development of specific probes for each case. However, there are several drawbacks to this approach: (1) it cannot differentiate between chromosomally normal embryos and balanced translocation carriers, and (2) chromosome probes for these regions are not always available because of the highly repetitive nature of these regions, resulting in cross-hybridization of probes (Scriven et al. 1998).

Application of FISH for chromosome translocations (Simpson 2010) in embryos (Munne et al. 1998) or polar bodies (Munne et al. 2000) became widespread in the late 1990s and into the 2000s. Numerous studies have reported successful pregnancies following PGD for translocations (Coonen et al. 2000; Chen et al. 2007; Lim et al. 2008; Wiland et al. 2008), and the strategy described above was rapidly adopted by IVF clinics worldwide.

To date, no randomized clinical trial has been performed for the use of FISH for chromosome translocations. Indeed, such trials are unlikely given that almost all translocations are family specific (with the exception of Robertsonian translocations). Notwithstanding the inability to determine normal from balanced translocations, the requirement for a bespoke couple-specific test, experimental errors in signal interpretation, and failure of probes to hybridize, the interphase FISH approach is still a robust one. FISH application for translocations has clearly shown a reduction in (1) spontaneous miscarriages, (2) abnormal viable offspring, and (3) the time needed to achieve a pregnancy, and it can be combined with aneuploidy screening (PGS) to assess the chromosomes not involved in the translocation itself (Simpson and Tempest 2010).

Despite the apparent robustness of the FISH approaches just described, array CGH (aCGH) approaches and single nucleotide polymorphism (SNP) microarrays are also being applied to translocation cases as they provide several advantages over interphase cytogenetics. First, they do not require lengthy, expensive patient-specific workups, and second, they provide genome-wide coverage and so can deliver aneuploidy data for all chromosomes, not just those involved in the translocation. In practical terms, there is no longer a requirement for a visible interphase nucleus at the time of embryo biopsy, which is required when using FISH on a single nucleus owing to the high chance of chromosome loss from a metaphase nucleus after nuclear spreading. Finally, SNP arrays should theoretically be able to distinguish between normal and balanced embryos (Simpson and Tempest 2010).

Assessment of Aneuploidy by Interphase Cytogenetics: The Rise, Fall, and Rise of PGS

The use of interphase cytogenetics for preventing X-linked disease transmission and exclusion of embryos with an unbalanced chromosomal complement paved the way for the use of FISH for chromosomal copy number in human preimplantation embryos and what subsequently evolved to become PGS.

It is estimated that the incidence of aneuploidy is approximately 0.6 % in newborns, 6 % in stillbirths, and 60 % in spontaneous abortions (Martin 2008). Most abnormalities are lethal and do not survive to term; however, certain chromosomal abnormalities do survive. With the advent of IVF and interphase cytogenetics in determining copy number of individual chromosomes in a PGD setting, it was possible to study the aneuploidy rates at this early stage of development (early studies reviewed by Griffin 1996). From the mid-1990s onward, the frequency of chromosome aneuploidies and estimates of chromosome mosaicism in the early human preimplantation embryo were reported (Munne et al. 1993, 1994, 1995). Munne et al. (2004) analyzed 2,000 embryos using probes for 14 chromosomes and found that the chromosomes most frequently involved in aneuploidy were 22, 16, 21, and 15, whereas the least involved were 14, X, and Y. They also reported higher rates of monosomy rather than trisomy. An important finding from studies in embryos was that the predominant type of mosaicism affecting preimplantation embryos was the diploid/aneuploid type arising from one of the first three divisions (probably first or second) (Delhanty et al. 1997; Daphnis et al. 2005). In a study by Daphnis et al. (2008), embryos were compared to investigate the evolution of chromosome abnormalities between the cleavage and blastocyst stages. The conclusion drawn was that a normal blastomere on day 3 is more likely to give rise to blastomeres with the correct chromosome complement on day 5, whereas an abnormal cell on day 3 predicts a poor outcome on day 5.

The studies on aneuploidy at the time were numerous. However, to summarize them, three trends emerged (Munne et al. 2007a):

1. Aneuploidy increases in cleavage-stage embryos with advancing maternal age, irrespective of embryo morphology.
2. Postmeiotic abnormalities (mosaicism, polyploidy, haploidy) increase with poorer embryo development and morphology.
3. Postmeiotic abnormalities are the most frequent type of abnormalities.

The main rationale for PGS is to increase the chance of a healthy pregnancy in subfertile patients undergoing IVF by screening for chromosome abnormalities, based on the fact that patients undergoing IVF with poor prognosis generate embryos with a high incidence of numerical chromosome abnormalities (60–70 %) (Donoso et al. 2007). Thus, theoretically, if abnormal embryos could be identified and excluded and chromosomally normal ones selected for embryo transfer, an improved pregnancy outcome (manifested as either reduced miscarriage rate, increased clinical pregnancy and live birth rate, increased implantation rate, shorter time to pregnancy, or a combination of these outcomes) should be expected, at least in women with a high risk of chromosome aneuploidy (Fritz 2008).

With this rationale in mind the main referral categories for PGS were advanced maternal age (AMA, more than 37 or 38 years), repeated implantation failure (RIF, three or more failed implantation attempts following embryo transfer), repeated miscarriage (RM, at least three pregnancy losses in patients with a normal karyotype), or severe male factor (SMF, abnormal semen parameters) (Donoso et al. 2007; Munne 2003; Harper et al. 2010).

The selection of probes predominantly used in the clinics was based on the incidence of chromosome abnormalities in spontaneous abortions and live births. Chromosomes 13, 15, 16, 18, 21, 22, X, and Y were the most widely used, enabling the detection of an estimated 72 % of abnormalities found in spontaneous abortions (Donoso et al. 2007). Over the years, the application of PGS had become widespread with 75 % of the overall PGD-related procedures in the United States, and 65 % in Europe, being PGS (Hernandez 2009). In the most recent ESHRE PGD consortium report, this steady increase has been highlighted with 3,401 PGS cycles in 2008 alone compared to 116 in 1997–1998 (Goossens et al. 2009, 2012).

Early studies using PGS reported an increase in implantation rates and, at the same time, a reduction in trisomic offspring and spontaneous abortions (Munne 2003; Harper et al. 2008; Werlin et al. 2003). However, criticisms of these early reports focus on the fact that these were nonrandomized and had poor experimental design, inadequate control groups, few or no reports on delivery rates, and relatively small patient numbers (Harper et al. 2008). The first randomized controlled trial (RCT) casting some doubt on the efficacy of PGS was published by Staessen et al. (2004) in which no difference in embryo implantation and pregnancy rates was reported between control and PGS patients.

A subsequent RCT study initiating wide debate and controversy with regard to the efficacy of PGS, published by Mastenbroek et al. (2007), reported a significant *decrease* in pregnancy rates and live births following PGS in women of advanced maternal age. This study was however, heavily criticized (Cohen and Grifo 2007; Wilton 2007; Munne et al. 2007b, c; Simpson 2008; Sermondade and Mandelbaum 2009; Handyside and Thornhill 2007) on many levels including experimental design, biopsy procedure, number of embryos per patient, and exclusion of probes for chromosomes 15 and 22. However, it prompted more RCTs (Jansen et al. 2008; Twisk et al. 2008; Mersereau et al. 2008; Hardarson et al. 2008; Staessen et al. 2008; Debrock et al. 2010; Chiamchanya et al. 2008; Schoolcraft et al. 2009; Garrisi et al. 2009), and a subsequent meta-analysis of the RCTs by Checa et al. (2009) concluded that, in women with poor prognosis or undergoing IVF, aneuploidy screening by PGS using FISH methodologies at the cleavage stage of development is associated with lower pregnancy and live birth rates. The reasons why PGS has failed to show a positive outcome in the RCTs have been extensively discussed elsewhere (Donoso et al. 2007; Beyer et al. 2009; Cohen et al. 2009; Wilton et al. 2009; Uher et al. 2009) and therefore are not considered further here.

The net effect of all this controversy is that FISH-based technologies have largely been superseded by array-based platforms for PGS. As this is a chapter on interphase cytogenetics, it is not within the remit of this chapter to discuss microarray-based approaches to detecting chromosome copy number. Suffice to say, however, that a brief flirtation with metaphase CGH (which uses a hybridization approach directly onto 'normal' human metaphase chromosomes, was rapidly replaced by aCGH, in which sample DNA is hybridized to glass slides containing small portions of chromosomes organized in precise locations or arrays). This approach is now finding widespread application with favorable outcomes for patients (Le Caignec et al. 2006; Fragouli et al. 2010, 2011; Schoolcraft et al. 2010) with successful live births reported (Fishel et al. 2010; Obradors et al. 2008).

In addition to aCGH, SNP arrays have been adapted for use in PGS (Brezina et al. 2011; Handyside et al. 2010; Treff et al. 2010, 2011; Northrop et al. 2010). SNP arrays take advantage of the sequence variation within a population, and this information can be used to distinguish one person from another or one chromosome from another in any person, because SNPs are biallelic and exist in either of two forms and occur in large numbers throughout the entire genome. Using information from parental DNA, SNP arrays can further extend their applicability in PGS by determining the parent and the meiotic phase of origin of the abnormality (Handyside et al. 2010; Gabriel et al. 2011). Further validation of these new platforms and well-designed RCTs to show improvement in delivery dates are under way to determine whether these approaches will help patients undergoing assisted reproductive technologies around the world (Harper an. Harton 2010).

The Future of Interphase Cytogenetics in Early Human Preimplantation Embryos

Although, in terms of PGS, FISH will soon be replaced by array-based methods and perhaps whole-genome sequencing, FISH remains a valid molecular cytogenetic tool for research purposes ostensibly because it is a relatively simple, robust, and low-cost method to determine chromosome copy number in individual nuclei. The cost alone of studies requiring analysis of thousands of individual cells makes the use of aCGH and SNP array approaches prohibitive. Using interphase cytogenetics for as many chromosomes as possible in preimplantation embryos will make it possible to gain insight into the types of abnormalities occurring (e.g., monosomies, trisomies), the level of mosaicism from analysis of all chromosomes and multiple cells from the same embryo, and the relationship between nuclear organization and chromosome abnormalities (discussed further next). Furthermore, in the next few years it could be used as a means of comparing the new array platforms for concordance purposes. Supportive evidence, which already exists in the literature, is discussed next.

Interphase cytogenetics is already in use for the validation of array-based approaches; this initially involved the use of nine probes in a two-layer experiment (Thornhill et al. 2005). Screening additional chromosomes, however, could provide a more comprehensive diagnosis. The added value of aneuploidy detection by screening 15 chromosomes in cryopreserved day 4 and 5 embryos using three rounds of hybridization is clear. Incorporation of an additional 6 chromosomes allowed detection of chromosome aberrations that were mainly mitotic in origin, leading to a higher percentage of mosaic embryos (Baart et al. 2007). In another study FISH was used to investigate chromosomal constitution of whole nontransferred embryos following the diagnosis of abnormality in a single biopsied blastomere from day 3 (DeUgarte et al. 2008). Of 198 abnormal embryos, 164 were confirmed when the whole embryo was analyzed by FISH, giving a positive predictive value of 83 %, signifying that 17 % of embryos are misdiagnosed as abnormal on day 3 when they are in fact normal (DeUgarte et al. 2008). Such follow-up FISH analysis of whole embryos provides another means of confirming the high level of mosaicism observed in cleavage-stage embryos and thus insight into how representative is a single cell with regard to the whole embryo. Using three rounds of hybridization and a mix of centromeric and telomeric probes in the last round, Colls et al. (2009) screened for 12 chromosomes and found that embryos diagnosed as normal for the initial chromosome panel (9 chromosomes) had extra abnormalities that would not have been found without extended screening. They postulated, however, that because of the use of telomeric probes in the later hybridization rounds and suboptimal conditions, the error rate was slightly higher from the percentage found for the 9 chromosomes alone (Colls et al. 2009). Thus, the extended screening can be important in revealing other 'rarer' chromosomal abnormalities found in preimplantation embryos and can also be tailored to different subgroups of patients.

Recently a 24-color FISH screen assay using four layers of hybridization and involving six fluorochromes per layer was described (Ioannou et al. 2011a). The methodology was applied successfully to preimplantation embryos not eligible for transfer to assess copy number and nuclear organization (Ioannou et al. 2012). Such an approach could be used to investigate if there are chromosome-specific mechanisms of error in early embryos and also used on blastocysts to assess aneuploidy rates on a cell-by-cell basis and compare data with ongoing pregnancies (Ioannou et al. 2012). It should be noted that, despite its obvious potential, if each probe fails (or the signal overlaps with its homologue) in 2 % of nuclei, a 24-chromosome screen would only be completely accurate in 40–50 % of nuclei, and this should be taken into account. Additional information obtained regarding the nuclear organization could provide further insight about its role in early human development, as described next (Ioannou and Griffin 2011).

Studying Nuclear Organization in Human Embryos

The rationale for the study of the nonrandom position of chromosomes in the interphase nuclei is well justified in other chapters in this book (Tanabe et al. 2001; Parada and Misteli 2002; Foster and Bridger 2005; Foster et al. 2005; Cremer and Cremer 2001; Manuelidis 1985, 1990; Oliver and Misteli 2005; Khalil et al. 2007; Meaburn and Misteli 2007; Meaburn et al. 2005). The location of chromosomes in the nucleus, its relationship to the accessibility of various nuclear machinery, and its vital role in the regulation of gene expression, DNA replication, damage, repair, development, and other cellular functions are well documented (Foster and Bridger 2005; Cremer and Cremer 2001; Miguel and Pombox 2006; Fraser and Bickmore 2007; Lanctot et al. 2007; Rouquette et al. 2010; Schoenfelder et al. 2010; Pederson 2004, 2011; Rajapakse and Groudine 2011; Dundr and Misteli 2011; Spector and Lamond 2011). The availability of multicolor images primarily used to detect chromosome copy number can provide an insight into nuclear organization and its role in early human development and the genesis of aneuploidy.

Nuclear organization in embryos is a relatively underexplored area mainly as a result of the issues outlined in the Introduction. Studying nuclear organization in human embryos is nonetheless important for a number of reasons. First, embryo cells are totipotent, and comparison of nuclear organization of these cells compared to committed cells in the body provides insight into the nature of totipotency itself. Second, between 50 and 70 % of human cleavage-stage blastomeres from IVF patients are chromosomally abnormal (Delhanty et al. 1997; Munne and Cohen 1998; Wells and Delhanty 2000; Voullaire et al. 2000), and one can speculate that nuclear localization is altered in aneuploid embryos compared to chromosomally normal ones. Finally, if specific patterns of nuclear organization do emerge, as already mentioned, they may provide novel markers for PGD and PGS.

To date, only a small number of studies have attempted to address the issue of nuclear organization in human embryos. Each of the studies used a basic computer-generated template with five concentric circles superimposed onto the two-dimensional (2D) nucleus before counting the number of FISH signals in each circle to record the chromosome position. The first investigated the position of seven chromosomes (13, 16, 18, 21, 22, X, and Y) in normal and abnormal human blastomeres using centromeric and locus-specific probes (McKenzie et al. 2004), finding that in normal blastomeres chromosome 13, 18, 21, and X were central and chromosome 16, 22, and Y were more peripherally located. However, this pattern was altered in aneuploid blastomeres with more peripheral localizations (McKenzie et al. 2004). The localization of specific chromosomes (13, 18, X) in the interior part of the nuclei during embryonic development, compared with the typical peripheral localization within a committed cell line, indicates a different nuclear organization pattern associated with totipotent cells at the cleavage stage of human development (McKenzie et al. 2004). A similar study by Diblik et al. (2007) reported that the localization of chromosomes 13, 16, 21, 22, X, and Y was essentially random in both normal and abnormal embryos. The only difference they could observe was that arrangement in chromosome 18 was significantly different to random distribution and shifted toward the periphery in aneuploid blastomeres (Diblik et al. 2007).

More recently, Finch et al. (2008) attempted to establish a correlation between chromosomal abnormalities and nuclear organization in human embryos and then compare this to a range of committed cell lines (Finch et al. 2008). This study reported a significant alteration of nuclear organization associated with chromosomally abnormal embryos compared to control committed cell lines. For example, chromosome 15 was localized in the periphery of nuclei in committed cells, but in aneuploid blastomeres chromosome 15 had a central localization. This study also reported that embryos with no detected abnormalities adopt a less distinct pattern in genome organization because of the existence of mixed populations of cells, each with a different nuclear organization (Finch et al. 2008). The foregoing studies suggest that examining nuclear architecture during early embryogenesis could provide insight into the mechanisms of aneuploidy and improve embryo selection in preimplantation diagnosis.

In the aforementioned studies, blastomeres were classified as normal based on ploidy from a subset of chromosomes: five chromosomes (13, 18, 21, X, Y) in the study by McKenzie et al. (2004), seven chromosomes (13, 16, 18, 21, 22, X, Y) in that by Diblik et al. (2007), and eight chromosomes (13, 15, 16, 18, 21, 22, X, Y) in the study by Finch et al. (2008).

Further investigations studied the nuclear localization of all 24 chromosomes using a methodology based on the 'mainstream' nuclear organization literature to analyze radial position (Ioannou et al. 2012) (Fig. 7.1). This approach allowed more accurate assessments of relative nuclear positions in blastomeres through analysis of the nuclear localization of all 24 chromosomes in 17 embryos (255 cells) and found that centromeres have a mainly central localization, giving evidence for a

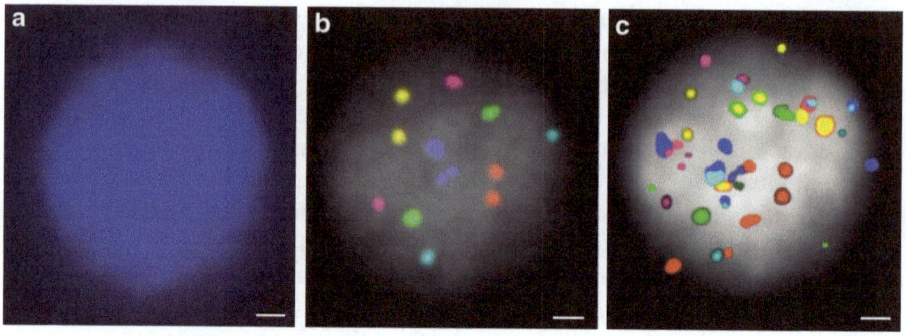

Fig. 7.1 Blastomere nucleus after four rounds of hybridization. *Bars* 10 μm. (**a**) DAPI only, showing retention of nuclear structure. (**b**) Same nucleus with final probe set signals shown: chromosome 19 in *blue*, chromosome 5 in *aqua*, chromosome 21 in *green*, chromosome 22 in *yellow*, chromosome 13 in *red*, chromosome 14, far-red fluorochrome pseudo-colored *purple*. (**c**) Same nucleus with probes from the other three rounds superimposed in Adobe Photoshop: note position and copy number of chromosomes 5, 13, 14, 19, 21, and 22 can still be observed (From Ioannou et al. 2012)

chromocenter nuclear organization as in sperm (Zalensky and Zalenskaya 2007; Ioannou et al. 2011b).

It is noteworthy that the various studies that have looked at nuclear organization of embryos have obtained quite variable results for the same chromosomes, possibly for numerous reasons, such as examination at different development stages, different embryo morphologies, different levels of aneuploidy, different patient cohorts, or technical reasons such as spreading techniques or analysis algorithms. Table 7.1 compares the results of the various studies.

The study of nuclear organization in human embryos is still in its infancy. Despite practical, ethical, and legal restrictions, some of the approaches outlined in this book may well be applied to human embryos if an appropriate clinical justification can be made. Studies of all chromosomes and their territories are possible as well as specific regions such as centromeres and telomeres. The issue of the relationships between aneuploidy, embryo morphology, maternal age, patient referral category, and so on all warrant further investigation. Such studies may well pave the way for investigations into human embryonic stem cells and the role of nuclear organization in stem cell differentiation.

Conclusions

Interphase cytogenetics in human embryos clearly has a number of applications both in the clinic and in the research laboratory. The other chapters outlined in this book pave the way for a range of exciting new studies that, potentially, could emerge on human embryos. Although interphase cytogenetics for PGS has already been superseded by array-based approaches, FISH is still a useful tool for the validation and

Table 7.1 Comparison of the nuclear organization studies for eight chromosome loci in human preimplantation embryos

References	**McKenzie *et al.* [103]**		**Diblik *et al.* [104]**		**Finch *et al.* [108]**		**Ioannou *et al.* [77]**	
Locus	**Normal**	**Abnormal**	**Normal**	**Abnormal**	**Normal**	**Abnormal**	**Normal**	**Abnormal**
13	Central	Peripheral	Random	Random	Random	Peripheral	Random	Central
15	N/A	N/A	N/A	N/A	Random	Central	Central	Central
16	Peripheral	Peripheral	Random	Random	Random	Central	Central /Medial	Central
18	Central	Peripheral	Central	Random	Random	Peripheral /Medial	Central	Central
21	Central	Peripheral	Random	Random	Random	Central	Central	Central
22	Peripheral	Peripheral	Random	Random	Random	Central	Central	Central
X	Central	Peripheral	Random	Random	Random	Random	Central	Central /Medial
Y	Peripheral	Peripheral	Random	Random	Random	Random	Central	Central

Note: The grey-shaded boxes denote concordance data between different studies for the Normal blastomeres. The red-shaded boxes denote concordance for two studies, although there is also a different concordance for the remaining two studies (e.g., 21, X). The green-shaded boxes denote concordance of data between different studies for the abnormal blastomeres 'follow-up' of clinical cases because of the rapid, low-cost, and robust cell-by-cell information that can be gleaned. Perhaps the most exciting avenue of research is that of nuclear organization. We still know very little about this in the earliest stages of our development when a unique set of cellular processes occur, not least of which is the establishment of our basic body plan.

References

Baart EB et al (2007) FISH analysis of 15 chromosomes in human day 4 and 5 preimplantation embryos: the added value of extended aneuploidy detection. Prenat Diagn 27(1):55–63

Beyer CE et al (2009) Preimplantation genetic screening outcomes are associated with culture conditions. Hum Reprod 24(5):1212–1220

Brezina PR, Brezina PR et al (2011) Single-gene testing combined with single nucleotide polymorphism microarray preimplantation genetic diagnosis for aneuploidy: a novel approach in optimizing pregnancy outcome. Fertil Steril 95(5):1786e5–1786e8

Chang LJ et al (2011) An update of preimplantation genetic diagnosis in gene diseases, chromosomal translocation, and aneuploidy screening. Clin Exp Reprod Med 38(3):126–134

Checa MA et al (2009) IVF/ICSI with or without preimplantation genetic screening for aneuploidy in couples without genetic disorders: a systematic review and meta-analysis. J Assist Reprod Genet 26(5):273–283

Chen Y et al (2007) A normal birth following preimplantation genetic diagnosis by FISH determination in the carriers of der(15)t(Y;15)(Yq12;15p11) translocations: two case reports. J Assist Reprod Genet 24(10):483–488

Chiamchanya C et al (2008) Preimplantation genetic screening (PGS) in infertile female age > or = 35 years by fluorescence in situ hybridization of chromosome 13, 18, 21, X and Y. J Med Assoc Thai 91(11):1644–1650

Cohen J, Grifo JA (2007) Multicentre trial of preimplantation genetic screening reported in the *New England Journal of Medicine*: an in-depth look at the findings. Reprod Biomed Online 15(4):365–366

Cohen J et al (2009) The role of preimplantation genetic diagnosis in diagnosing embryo aneuploidy. Curr Opin Obstet Gynecol 21(5):442–449

Colls P et al (2009) Increased efficiency of preimplantation genetic diagnosis for aneuploidy by testing 12 chromosomes. Reprod Biomed Online 19(4):532–538

Coonen E et al (2000) Preimplantation genetic diagnosis of a reciprocal translocation t(3;11)(q27.3;q24.3) in siblings. Mol Hum Reprod 6(3):199–206

Cremer T, Cremer C (2001) Chromosome territories, nuclear architecture and gene regulation in mammalian cells. Nat Rev Genet 2(4):292–301

Daphnis DD et al (2005) Detailed FISH analysis of day 5 human embryos reveals the mechanisms leading to mosaic aneuploidy. Hum Reprod 20(1):129–137

Daphnis DD et al (2008) Analysis of the evolution of chromosome abnormalities in human embryos from day 3 to 5 using CGH and FISH. Mol Hum Reprod 14(2):117–125

Debrock S et al (2010) Preimplantation genetic screening for aneuploidy of embryos after in vitro fertilization in women aged at least 35 years: a prospective randomized trial. Fertil Steril 93(2):364–373

Delhanty JD et al (1997) Multicolour FISH detects frequent chromosomal mosaicism and chaotic division in normal preimplantation embryos from fertile patients. Hum Genet 99(6):755–760

DeUgarte CM et al (2008) Accuracy of FISH analysis in predicting chromosomal status in patients undergoing preimplantation genetic diagnosis. Fertil Steril 90(4):1049–1054

Diblik J et al (2007) Chromosome topology in normal and aneuploid blastomeres from human embryos. Prenat Diagn 27(12):1091–1099

Donoso P et al (2007) Current value of preimplantation genetic aneuploidy screening in IVF. Hum Reprod Update 13(1):15–25
Dundr M, Misteli T (2011) Biogenesis of nuclear bodies. Cold Spring Harb Perspect Biol 2(12):a000711
Finch KA et al (2008) Nuclear organisation in totipotent human nuclei and its relationship to chromosomal abnormality. J Cell Sci 121(pt 5):655–663
Fishel S, Fishel S et al (2010) Live birth after polar body array comparative genomic hybridization prediction of embryo ploidy: the future of IVF? Fertil Steril 93(3):1006 e7–1006 e10
Foster HA, Bridger JM (2005) The genome and the nucleus: a marriage made by evolution. Genome organisation and nuclear architecture. Chromosoma (Berl) 114(4):212–229
Foster HA et al (2005) Non-random chromosome positioning in mammalian sperm nuclei, with migration of the sex chromosomes during late spermatogenesis. J Cell Sci 118(pt 9):1811–1820
Fragouli E et al (2010) Comprehensive chromosome screening of polar bodies and blastocysts from couples experiencing repeated implantation failure. Fertil Steril 94(3):875–887
Fragouli E et al (2011) Cytogenetic analysis of human blastocysts with the use of FISH, CGH and aCGH: scientific data and technical evaluation. Hum Reprod 26(2):480–490
Fraser P, Bickmore W (2007) Nuclear organization of the genome and the potential for gene regulation. Nature (Lond) 447(7143):413–417
Fritz MA (2008) Perspectives on the efficacy and indications for preimplantation genetic screening: where are we now? Hum Reprod 23(12):2617–2621
Gabriel AS et al (2011) An algorithm for determining the origin of trisomy and the positions of chiasmata from SNP genotype data. Chromosome Res 19(2):155–163
Garrisi JG et al (2009) Effect of infertility, maternal age, and number of previous miscarriages on the outcome of preimplantation genetic diagnosis for idiopathic recurrent pregnancy loss. Fertil Steril 92(1):288–295
Goossens V et al (2009) ESHRE PGD Consortium data collection IX: cycles from January to December 2006 with pregnancy follow-up to October 2007. Hum Reprod 24(8):1786–1810
Goossens V et al (2012) ESHRE PGD Consortium data collection XI: cycles from January to December 2008 with pregnancy follow-up to October 2009. Hum Reprod 27(7):1887–1911
Griffin DK (1994) Fluorescent in situ hybridization for the diagnosis of genetic disease at postnatal, prenatal, and preimplantation stages. Int Rev Cytol 153:1–40
Griffin DK (1996) The incidence, origin, and etiology of aneuploidy. Int Rev Cytol 167:263–296
Griffin DK et al (1991) Fluorescent in-situ hybridization to interphase nuclei of human preimplantation embryos with X and Y chromosome specific probes. Hum Reprod 6(1):101–105
Griffin DK et al (1992) Dual fluorescent in situ hybridisation for simultaneous detection of X and Y chromosome-specific probes for the sexing of human preimplantation embryonic nuclei. Hum Genet 89(1):18–22
Griffin DK et al (1993) Diagnosis of sex in preimplantation embryos by fluorescent in situ hybridisation. BMJ 306(6889):1382
Handyside A, Thornhill A (2007) In vitro fertilisation with preimplantation genetic screening. N Engl J Med 357(17):1770
Handyside AH et al (2010) Karyomapping: a universal method for genome wide analysis of genetic disease based on mapping crossovers between parental haplotypes. J Med Genet 47(10):651–658
Hardarson T et al (2008) Preimplantation genetic screening in women of advanced maternal age caused a decrease in clinical pregnancy rate: a randomized controlled trial. Hum Reprod 23(12):2806–2812
Harper JC, Harton G (2010) The use of arrays in preimplantation genetic diagnosis and screening. Fertil Steril 94(4):1173–1177
Harper JC et al (1994) Identification of the sex of human preimplantation embryos in two hours using an improved spreading method and fluorescent in-situ hybridization (FISH) using directly labelled probes. Hum Reprod 9(4):721–724

Harper J et al (2008) What next for preimplantation genetic screening? Hum Reprod 23(3):478–480
Harper J et al (2010) What next for preimplantation genetic screening (PGS)? A position statement from the ESHRE PGD Consortium steering committee. Hum Reprod 25(4):821–823
Hernandez ER (2009) What next for preimplantation genetic screening? Beyond aneuploidy. Hum Reprod 24(7):1538–1541
Ioannou D, Griffin DK (2011) Male fertility, chromosome abnormalities, and nuclear organization. Cytogenet Genome Res 133(2-4):269–279
Ioannou D et al (2011a) Multicolour interphase cytogenetics: 24 chromosome probes, 6 colours, 4 layers. Mol Cell Probes 25(5-6):199–205
Ioannou D et al (2011b) Nuclear organisation of sperm remains remarkably unaffected in the presence of defective spermatogenesis. Chromosome Res 19(6):741–753
Ioannou D et al (2012) Twenty-four chromosome FISH in human IVF embryos reveals patterns of post-zygotic chromosome segregation and nuclear organisation. Chromosome Res 20(4):447–460
Jansen RP et al (2008) What next for preimplantation genetic screening (PGS)? Experience with blastocyst biopsy and testing for aneuploidy. Hum Reprod 23(7):1476–1478
Khalil A et al (2007) Chromosome territories have a highly nonspherical morphology and nonrandom positioning. Chromosome Res 15(7):899–916
Lanctot C et al (2007) Dynamic genome architecture in the nuclear space: regulation of gene expression in three dimensions. Nat Rev Genet 8(2):104–115
Le Caignec C et al (2006) Single-cell chromosomal imbalances detection by array CGH. Nucleic Acids Res 34(9):e68
Lim CK et al (2008) A healthy live birth after successful preimplantation genetic diagnosis for carriers of complex chromosome rearrangements. Fertil Steril 90(5):1680–1684
Manuelidis L (1985) Individual interphase chromosome domains revealed by in situ hybridization. Hum Genet 71(4):288–293
Manuelidis L (1990) A view of interphase chromosomes. Science 250(4987):1533–1540
Martin RH (2008) Meiotic errors in human oogenesis and spermatogenesis. Reprod Biomed Online 16(4):523–531
Mastenbroek S et al (2007) In vitro fertilization with preimplantation genetic screening. N Engl J Med 357(1):9–17
McKenzie LJ et al (2004) Nuclear chromosomal localization in human preimplantation embryos: correlation with aneuploidy and embryo morphology. Hum Reprod 19(10):2231–2237
Meaburn KJ, Misteli T (2007) Cell biology: chromosome territories. Nature (Lond) 445(7126):379–781
Meaburn KJ, Parris CN, Bridger JM (2005) The manipulation of chromosomes by mankind: the uses of microcell-mediated chromosome transfer. Chromosoma (Berl) 114(4):263–274
Mersereau JE et al (2008) Preimplantation genetic screening to improve in vitro fertilization pregnancy rates: a prospective randomized controlled trial. Fertil Steril 90(4):1287–1289
Mewar R et al (1992) Confirmation of a cryptic unbalanced translocation using whole chromosome fluorescence in situ hybridization. Am J Med Genet 44(4):477–481
Miguel RB, Pombox A (2006) Intermingling of chromosome territories in interphase suggests role in translocations and transcription-dependent associations. PLoS Biol 4:780–788
Munne S (2003) Preimplantation genetic diagnosis and human implantation—a review. Placenta 24(suppl B):S70–S76
Munne S, Cohen J (1998) Chromosome abnormalities in human embryos. Hum Reprod Update 4(6):842–855
Munne S et al (1993) Diagnosis of major chromosome aneuploidies in human preimplantation embryos. Hum Reprod 8(12):2185–2191
Munne S et al (1994) Chromosome mosaicism in human embryos. Biol Reprod 51(3):373–379
Munne S, et al (1995) Assessment of numeric abnormalities of X, Y, 18, and 16 chromosomes in preimplantation human embryos before transfer. Am J Obstet Gynecol 172(4 pt 1):1191–1199; discussion 1199–1201

Munne S et al (1998) Preimplantation genetic analysis of translocations: case-specific probes for interphase cell analysis. Hum Genet 102(6):663–674
Munne S et al (2000) Outcome of preimplantation genetic diagnosis of translocations. Fertil Steril 73(6):1209–1218
Munne S et al (2004) Differences in chromosome susceptibility to aneuploidy and survival to first trimester. Reprod Biomed Online 8(1):81–90
Munne S et al (2007a) Maternal age, morphology, development and chromosome abnormalities in over 6000 cleavage-stage embryos. Reprod Biomed Online 14(5):628–634
Munne S et al (2007b) Substandard application of preimplantation genetic screening may interfere with its clinical success. Fertil Steril 88(4):781–784
Munne S, Cohen J, Simpson JL (2007c) In vitro fertilization with preimplantation genetic screening. N Engl J Med 357(17):1769–1770
Northrop LE et al (2010) SNP microarray-based 24 chromosome aneuploidy screening demonstrates that cleavage-stage FISH poorly predicts aneuploidy in embryos that develop to morphologically normal blastocysts. Mol Hum Reprod 16(8):590–600
Obradors A et al (2008) Birth of a healthy boy after a double factor PGD in a couple carrying a genetic disease and at risk for aneuploidy: case report. Hum Reprod 23(8):1949–1956
Oliver B, Misteli T (2005) A non-random walk through the genome. Genome Biol 6(4):214
Parada L, Misteli T (2002) Chromosome positioning in the interphase nucleus. Trends Cell Biol 12(9):425–432
Pederson T (2004) The spatial organization of the genome in mammalian cells. Curr Opin Genet Dev 14(2):203–209
Pederson T (2011) The nucleolus. Cold Spring Harb Perspect Biol 3(3):a000638
Rajapakse I, Groudine M (2011) On emerging nuclear order. J Cell Biol 192(5):711–721
Rouquette J et al (2010) Functional nuclear architecture studied by microscopy: present and future. Int Rev Cell Mol Biol 282:1–90
Schoenfelder S, Clay I, Fraser P (2010) The transcriptional interactome: gene expression in 3D. Curr Opin Genet Dev 20(2):127–133
Schoolcraft WB et al (2009) Preimplantation aneuploidy testing for infertile patients of advanced maternal age: a randomized prospective trial. Fertil Steril 92(1):157–162
Schoolcraft WB et al (2010) Clinical application of comprehensive chromosomal screening at the blastocyst stage. Fertil Steril 94(5):1700–1706
Scriven PN, Handyside AH, Ogilvie CM (1998) Chromosome translocations: segregation modes and strategies for preimplantation genetic diagnosis. Prenat Diagn 18(13):1437–1449
Sermondade N, Mandelbaum J (2009) [Mastenbroek controversy or how much ink is spilled on preimplantation genetic screening subject]. Gynecol Obstet Fertil 37(3):252–256
Simpson JL (2008) What next for preimplantation genetic screening? Randomized clinical trial in assessing PGS: necessary but not sufficient. Hum Reprod 23(10):2179–2181
Simpson JL (2010) Preimplantation genetic diagnosis at 20 years. Prenat Diagn 30(7):682–695
Simpson J, Tempest H (2010) Role of preimplantation genetic diagnosis (PGD) in current infertility practice. Int J Infertility Fetal Med 1(1):1–10
Spector DL, Lamond AI (2011) Nuclear speckles. Cold Spring Harb Perspect Biol 3(2):a000646
Staessen C et al (2004) Comparison of blastocyst transfer with or without preimplantation genetic diagnosis for aneuploidy screening in couples with advanced maternal age: a prospective randomized controlled trial. Hum Reprod 19(12):2849–2858
Staessen C et al (2008) Preimplantation genetic screening does not improve delivery rate in women under the age of 36 following single-embryo transfer. Hum Reprod 23(12):2818–2825
Tanabe H et al (2001) Non-random radial arrangements of interphase chromosome territories: evolutionary considerations and functional implications. Mutat Res 504:37–45
Thornhill AR et al (2005) ESHRE PGD Consortium 'Best practice guidelines for clinical preimplantation genetic diagnosis (PGD) and preimplantation genetic screening (PGS)'. Hum Reprod 20(1):35–48
Treff NR et al (2010) SNP microarray-based 24 chromosome aneuploidy screening is significantly more consistent than FISH. Mol Hum Reprod 16(8):583–589

Treff NR et al (2011) Single-cell whole-genome amplification technique impacts the accuracy of SNP microarray-based genotyping and copy number analyses. Mol Hum Reprod 17(6): 335–343

Twisk M et al (2008) No beneficial effect of preimplantation genetic screening in women of advanced maternal age with a high risk for embryonic aneuploidy. Hum Reprod 23(12): 2813–2817

Uher P et al (2009) Non-informative results and monosomies in PGD: the importance of a third round of re-hybridization. Reprod Biomed Online 19(4):539–546

Voullaire L et al (2000) Chromosome analysis of blastomeres from human embryos by using comparative genomic hybridization. Hum Genet 106(2):210–217

Wells D, Delhanty JD (2000) Comprehensive chromosomal analysis of human preimplantation embryos using whole genome amplification and single cell comparative genomic hybridization. Mol Hum Reprod 6(11):1055–1062

Werlin L et al (2003) Preimplantation genetic diagnosis as both a therapeutic and diagnostic tool in assisted reproductive technology. Fertil Steril 80(2):467–468

Wiland E et al (2008) Successful pregnancy after preimplantation genetic diagnosis for carrier of t(2;7)(p11.2;q22) with high rates of unbalanced sperm and embryos: a case report. Prenat Diagn 28(1):36–41

Wilton LJ (2007) In vitro fertilization with preimplantation genetic screening. N Engl J Med 357(17):1770; author reply 1770–1771

Wilton L et al (2009) The causes of misdiagnosis and adverse outcomes in PGD. Hum Reprod 24(5):1221–1228

Zalensky A, Zalenskaya I (2007) Organization of chromosomes in spermatozoa: an additional layer of epigenetic information? Biochem Soc Trans 35(pt 3):609–611

Chapter 8
Organization of Chromosomes in Human Sperm Nucleus

Olga S. Mudrak, Liudmila V. Solovjeva, and Vadim O. Chagin

Abstract Chromosomes in the human sperm nucleus adopt a hierarchy of structures starting from protamine toroids, the elementary units of DNA packaging, and up to the higher-order organization of chromosome territories. Nonrandom intranuclear positioning of individual chromosomes, with centromeres aggregated in the internally located chromocenter and preferential location of telomeres at the nuclear periphery, suggests a highly defined architecture of the sperm nucleus at the level of genome. Such an ordered chromatin organization in the sperm nucleus may have a functional significance and determine the onset of paternal gene activity at early stages of embryonic development. This chapter describes relevant experimental data with primary attention to studies of human spermatozoa and discusses the implications of sperm chromosome organization for male reproductive health.

Abbreviations

CHR	Chromosome
CT	Chromosome territory
FISH	Fluorescence in situ hybridization
ICSI	Intracytoplasmic sperm injection
MSCI	Meiotic sex chromosome inactivation
MSUC	Meiotic silencing of unsynapsed chromatin
WCP	Whole chromosome painting probe

O.S. Mudrak (✉) • L.V. Solovjeva • V.O. Chagin
Institute of Cytology, Russian Academy of Sciences, Tikhoretsky Avenue 4,
St. Petersburg 194064, Russia
e-mail: olgamudrak2010@gmail.com; Mila.Solovjeva@gmail.com; dr.chagin@gmail.com

Y.B. Yurov et al. (eds.), *Human Interphase Chromosomes: Biomedical Aspects*,
DOI 10.1007/978-1-4614-6558-4_8,

Introduction

Our understanding of higher-order chromatin arrangement in somatic cell nucleus (referred to as nuclear architecture, genome architecture, or chromosome architecture) has progressed considerably in recent decades, mainly as the result of the development of a wide variety of hybridization probes for specific genomic sequences and advances in fluorescence imaging techniques. It has been established that chromosomes are compartmentalized into distinct subvolumes—chromosome territories—that are nonrandomly distributed in the nucleus. The organization of chromosome territories and chromosome domains, chromatin structure, and their dynamics inside the interphase nucleus have functional consequences for gene expression regulation (reviewed in Cremer and Cremer 2001; Lanctôt et al. 2007; Misteli 2007; Pai and Engelke 2010; Meister et al. 2011).

Spermatozoa are the specialized cells that differentiate from spermatogonia precursor cells in the course of spermatogenesis. Spermatogenesis includes two major stages: meiosis and spermiogenesis. Spermatogonia are divided mitotically and differentiate into spermatocytes that undergo meiotic divisions which result in formation of haploid spermatids. During spermiogenesis, the maturation of spermatids into motile spermatozoa occurs, which is accompanied by the formation of the acrosome, the loss of the excess cytoplasm, and the formation of a flagellum. The maturation is also characterized by chromatin reorganization: the replacement of somatic histones with sperm-specific proteins (protamines), supercondensation of DNA, and inhibition of transcriptional activity. For a long time, the compact hydrodynamically shaped sperm cell with inert DNA was considered as a "sack" of genes unlikely to bear any information beyond its genomic load. However, a complex nonrandom organization of the genome was demonstrated for mammalian sperm cells (reviewed by Mudrak et al. 2011). An emerging concept is that in addition to the paternal genome per se, the sperm nucleus provides a structural context that is essential for proper genome activation upon fertilization.

This chapter summarizes the current state of knowledge on human sperm chromosome organization at different structural levels, starting from the packaging of the DNA with protamines to the higher-order chromosome configuration and intranuclear chromosome positioning. Possible implications of sperm chromosome architecture for male factor infertility are discussed.

DNA Packaging in Spermatozoa

The chromatin of the sperm nucleus is characterized by extensive protamination that leads to the tight packaging of chromatin, which is necessary for the protection of parental genome during movement of the spermatozoa through the female reproductive tract.

There are two types of protamines in humans, P1 and P2. The P1/P2 ratio ranges from 0.54 to 1.43 in healthy individuals (Nanassy et al. 2011). Protamines are highly basic proteins that are characterized by an arginine-rich core and cysteine residues.

Positively charged arginine residues electrostatically neutralize DNA, leading to a strong protamine–DNA binding. The cysteine residues of protamines mediate the formation of multiple inter- and intramolecular disulfide bonds so that sperm DNA is organized into a nucleoprotamine complex with extreme compactness (Balhorn 1982, 2007).

Protamines coil sperm DNA into toroidal structures approximately 50–70 nm in outer diameter and 25 nm thick, a form of packaging that inactivates the sperm genome (Brewer et al. 1999; Balhorn 2007). Each toroid contains 50 kb DNA and is linked to the neighboring ones by nuclease-sensitive histone-bound DNA fragments, toroid linkers (Sotolongo et al. 2005; Ward 2010). Stacking of these nucleoprotamine toroids results in DNA compaction approximately tenfold more efficient as compared with nucleohistone packaging. Toroids and core particles remain in decondensed human sperm nuclei, as was demonstrated by the method of atomic force microscopy (Hud et al. 1993; Allen et al. 1996; Joshi et al. 2000; Nazarov et al. 2008).

According to the "donut-loop" model of chromatin organization in mammalian sperm suggested by S.W. Ward (reviewed by Ward and Ward 2004), during spermiogenesis, individual DNA loop domains condense into toroidal (donut-like) structures, so that each toroid contains a single DNA loop domain that is attached to the sperm nuclear matrix via toroid linkers (Ward 2010).

Mudrak et al. (2005) reported 500-nm chromatin beads in the partially decondensed human sperm nuclei and suggested that they consisted of a number of toroid stacks, thus representing the next hierarchical level of sperm chromatin organization. Earlier, Haaf and Ward, using fluorescence in situ hybridization (FISH) on sperm chromatin spreads, observed beaded fibers with basic package units of 180, 360, and 600 nm (Haaf and Ward 1995). The relationship of such beads to nucleoprotamine toroids remains to be elucidated. At the next level, two rows of 500-nm chromatin beads interconnected by thinner and less dense chromatin filaments organize the chromosome-arm fiber with thickness about 1,000 nm (Mudrak et al. 2005).

It has been demonstrated that replacement of somatic histones by protamines during spermiogenesis occurs in a stepwise manner. First (during meiosis), somatic histones are replaced with testis-specific histone variants; then (during chromatin condensation at the elongating spermatid stage of spermiogenesis), the latter are replaced by transition proteins, which, in turn, are replaced by protamines (Meistrich et al. 2003; Churikov et al. 2004; reviewed in Mudrak et al. 2011). As a result, in mature spermatozoa, histones are not completely replaced by protamines; 1–2 % of mouse sperm DNA (Balhorn 1982; Pittoggi et al. 1999; Brykczynska et al. 2010) and as much as 15 % of human sperm DNA remains histone bound and packed into nucleosomes (Churikov et al. 2004; Gineitis et al. 2000; Hammoud et al. 2009).

Mature human spermatozoa contain core histones (H2A, H2B, H3, H4) (Gatewood et al. 1990); histone variants such as CENP-A (Zalensky et al. 1993), TSH2B (Zalensky et al. 2002); H3.1, 3.2, 3.3 (van der Heijden et al. 2008), H2A.Z (Hammoud et al. 2009), and H2AX (Li et al. 2006); and the modified histones H3K4 Me2, Me3, and H3K9 Me3 (Hammoud et al. 2009).

Histone hTSH2B (human testis/sperm-specific H2B variant), which is transcribed exclusively in the testis, is also found in spermatogonia, spermatids, and in a subpopulation of mature sperm cells (Zalensky et al. 2002), where it is involved in nuclear decondensation during fertilization (Singleton et al. 2007).

Phosphosphorylated histone H2AX (γH2AX) plays a crucial role in the process known as meiotic sex chromosome inactivation (MSCI), a form of X-chromosome inactivation present in male germ cells. γH2AX marks entire sex chromosomes at the onset of MSCI (Mahadevaiah et al. 2001).

H3.1/H3.2 histone variants originated from male germ cells were found in paternal pronuclei in pre-S-phase human zygotes (van der Heijden et al. 2008).

Several research groups addressed the question whether histone-bound DNA distributed randomly throughout the sperm genome or the histones were associated with certain sequences. Earlier work by Gatewood et al. (1987) suggested that histone binding was restricted to specific sequences (Gatewood et al. 1987). Later, analysis of the chromatin structure of members of β-globin gene family expressed at different times during development showed that embryo-specific ε- and γ-globin genes were histone enriched, whereas the postnatal-expressed β-globin gene was protamine enriched in human sperm cells (Gardiner-Garden et al. 1998). It was suggested that histones in sperm chromatin might mark sets of genes that would be preferentially activated during early development. It was shown in subsequent studies that histone-associated DNA included telomeres (Zalenskaya et al. 2000), centromeres (Zalensky et al. 1993; Hammoud et al. 2009), paternally imprinted regions (Banerjee and Smallwood 1995; Hammoud et al. 2009), specific gene loci (Wykes and Krawetz 2003; Hammoud et al. 2009), and regulatory sequences of developmentally important genes (Arpanahi et al. 2009). Recent data from Arpanahi et al. (2009) and Hammoud et al. (2009) have demonstrated that histone-bound DNA is distributed in paternal genome in two ways: in relatively large tracts, ranging from 10 to 100 kb, and in smaller tracts interspersed throughout the genome. Smaller tracts of histone-associated DNA distributed evenly throughout the paternal genome were suggested to represent a repeating unit of sperm chromatin structure, such as linker regions between protamine toroids in the chromatin fiber (Ward 2010).

Together, these findings suggest that there are at least two types of sperm chromatin: nucleoprotamine-containing domains and nucleohistone-containing domains. The nucleohistone domains are structurally more open as compared to the nucleoprotamine domains and are nuclease sensitive in contrast to nuclease-resistant nucleoprotamine domains.

Histone-associated chromatin might represent the residual active chromatin that persists through chromatin condensation and therefore reveals the transcriptional history of spermatogenesis (Johnson et al. 2011).

The transmission of core histones together with modified histones and histone variants to the egg leaves the door open for DNA and histone-based epigenetic signals that may be important for consequent embryonic development (Miller et al. 2010).

It has been speculated that sperm genome is frozen in a dynamic configuration that reflects its pending introduction to the ooplasm (Miller et al. 2010). Deviation from this configuration caused by DNA-packaging anomalies may lead to infertile phenotypes. It has been shown that abnormally high P1/P2 ratios are associated with decreased fertilization rates and embryo quality (Simon et al. 2011). In line with this observation, an aberrant expression of P2 induced by oxidative stress caused by cigarette smoking leads to abnormal elevation of the P1/P2 level and improper sperm chromatin condensation (Hammadeh et al. 2010). Antioxidant therapy after oxidative stress resulted in significant improvements in sperm chromatin integrity and protamine packaging (Tunc et al. 2009).

Defects in DNA methylation at imprinted loci have been found in oligozoospermic patients and patients with improper histone replacement by protamines, indicating the existence of a tight link between epigenetic chromatin alterations and male infertility (Carrell and Hammoud 2010; Hammoud et al. 2010).

Therefore, similar to somatic cells, the structural organization of chromatin in the sperm nucleus apparently has a functional significance, and its alterations can affect spermatogenesis and early developmental processes.

Higher-Order Chromatin Organization in the Sperm Nucleus

In the following sections we describe elements of sperm genome architecture above the level of nucleoprotamine/nucleohistone complexes, revealed by methods of FISH and microscopy.

Application of FISH to Chromosome Positioning Studies in Spermatozoa

Almost 40 years ago, Geraedts and Pearson (1975) demonstrated that in human spermatozoa, chromosome 1 (CHR 1) is frequently located adjacent to CHR Y, which suggested a nonrandom chromosome arrangement in the human sperm nucleus. Later, Luetjens et al. (1999) showed that CHR 18 had a tendency to locate in the basal area of the nucleus, near the sperm tail, whereas CHR X was preferentially located in the apical area (Luetjens et al. 1999).

FISH studies using whole chromosome painting (WCP) probes demonstrated that, similarly to somatic cells, individual chromosomes occupy distinct territories in human spermatozoa (Haaf and Ward 1995; Zalensky et al. 1995; Hazzouri et al. 2000). In the following years, data on the defined chromosome positioning in human sperm began to appear (Hazzouri et al. 2000, Tilgen et al. 2001; Sbracia et al. 2002; Zalenskaya and Zalensky 2004; Mudrak et al. 2005; Wiland et al. 2008; Manvelyan et al. 2008). These studies provided a strong basis for the current view of nonrandom chromosome arrangement in human spermatozoa.

By now, information on the location of all 23 chromosomes in the haploid human male gamete is available. However, it is not entirely conclusive, because application of different protocols resulted in inconsistent data on chromosome positioning coming from different laboratories.

Centromeres are traditionally used in somatic cells as indicators of intranuclear chromosome location (Nagele et al. 1999; Sun and Yokota 1999). Along with WCP, chromosome-specific centromeric probes were employed in chromosome positioning studies on sperm cells (Sbracia et al. 2002; Zalenskaya and Zalensky 2004; Finch et al. 2008; Wiland et al. 2008). The WCP probe, covering the entire length of the chromosome, yields the large FISH signal; the intranuclear position of the chromosome territory (CT), in this case, is often determined by the position of the hybridization signal center (Foster et al. 2005).

FISH studies on sperm cells have some peculiarities originating from sperm cell characteristics. For example, the mature spermatozoon has almost no cytoplasm. Its nucleus occupies almost the whole of the sperm head. The amount of cytoplasm is diminished during spermiogenesis to reduce the cell weight and achieve a more hydrodynamic shape, improving sperm motility in the female reproductive tract. Therefore, no pretreatment is needed to remove the cytoplasm before denaturation to enable efficient hybridization of the probe to the target.

Although removal of cytoplasmic debris can be omitted, the decondensation of densely packed sperm chromatin is a prerequisite for performing FISH. Because of its extensive protamination, sperm DNA is the most condensed DNA in eukaryotes. Therefore, to enable penetration of painting probes and antibodies during FISH, sperm chromatin needs to be relaxed. Sperm chromatin decondensing buffers slightly differ from each other, but all include dithiothreitol (DTT), the thiol reducing agent, which reduces disulfide bonds between protamines (Zalensky et al. 1995, 1997; Luetjens et al. 1999; Hazzouri et al. 2000; Zalenskaya and Zalensky 2004; Mudrak et al. 2005; Finch et al. 2008; Wiland et al. 2008).

Longitudinal and Radial Chromosome Positioning in Spermatozoa

Asymmetry of the sperm nucleus and the presence of a spatial reference point, the place of tail attachment, facilitate chromosome positioning studies. The elongated shape of the human sperm nucleus allows the assessment of not only radial but also longitudinal chromosome position.

To determine the longitudinal position, that is, the location of the chromosome along the anterior–posterior axis, the length of the sperm nucleus can be divided into equal sectors: "subacrosomal, equatorial, basal" (Sbracia et al. 2002), "regions (I–IV)" (Zalenskaya and Zalensky 2004), and "apical, medial, basal" (Wiland et al. 2008), and the number of FISH signals in each sector is calculated. Alternatively, the normalized distance from the center of the FISH signal to the tail attachment spot can be computed and used as the indicator of longitudinal chromosome location (Zalenskaya and Zalensky 2004).

The preferential location in the anterior part of the human sperm nucleus has been demonstrated for CHR X (Luetjens et al. 1999; Hazzouri et al. 2000; Sbracia et al. 2002; Zalenskaya and Zalensky 2004) and CHR 1 (Zalenskaya and Zalensky 2004; Mudrak et al. 2005), whereas CHR 18 has been found preferentially in the posterior part (Luetjens et al. 1999; Sbracia et al. 2002; Mudrak et al. 2012). A random distribution along the long nuclear axis has been demonstrated for CHR 13 (Hazzouri et al. 2000).

Radial position can be defined as the position of chromosomes in relationship to the nuclear periphery or the nuclear interior. Some authors divided sperm nuclei into concentric zones, "central and peripheral" (Manvelyan et al. 2008), or enumerated "shells" (Finch et al. 2008), and scored the number of FISH signals in each zone. Other authors ascertained the distance from FISH signals either to the nearest peripheral edge (Foster et al. 2005) or to the long axis of the nucleus (Zalenskaya and Zalensky 2004). It has been shown that CHR 6 (Zalenskaya and Zalensky 2004), CHR 7 (Manvelyan et al. 2008), and CHR 18 (Manvelyan et al. 2008; Mudrak et al. 2012) are mostly peripheral in location, whereas CHR X (Zalenskaya and Zalensky 2004; Mudrak et al. 2012) and CHR 19 (Manvelyan et al. 2008; Mudrak et al. 2012) are internal. Figure 8.1a–c demonstrates different approaches to chromosome position determination. Some disagreements in data on chromosome localization between different research groups can result from different methods of cell treatment and FISH data analysis. For instance, the internal location of CHR X and peripheral location of CHR 6 (Zalenskaya and Zalensky 2004) were reported, as compared to the peripheral position of CHR X and central position of CHR 6 found in another study (Manvelyan et al. 2008).

Analysis of three-dimensional (3D) chromosome arrangement in human spermatozoa (Manvelyan et al. 2008) has shown the correlation of radial positioning with chromosome size and gene content. Large chromosomes occupied mostly the periphery and small chromosomes the nuclear interior: gene-rich chromosomes were more centrally located in comparison with gene-poor chromosomes.

A similar tendency in the radial intranuclear positioning of chromosomes of different size and gene content was observed in human somatic cells (Croft et al. 1999; Bickmore and Chubb 2003; Parada et al. 2004). However, detailed study of chromosome positioning in spermatozoa from three breeds of pig (Foster et al. 2005), although showing the association between position and gene density, did not demonstrate any correlation between position and chromosome size.

Interestingly, a nonrandom chromosome arrangement has been reported in sperm cells of other mammals: rat (Meyer-Ficca et al. 1998), mouse (Garagna et al. 2001), pig (Foster et al. 2005), cow (Mudrak et al., unpublished data), and mammals of the earliest groups, such as marsupials (Greaves et al. 2001) and monotremes (Watson et al. 1996; Greaves et al. 2003; Tsend-Ayush et al. 2009).

The observed evolutionary conservation of nonrandom chromosome arrangement in mammalian sperm cells implies its functional importance.

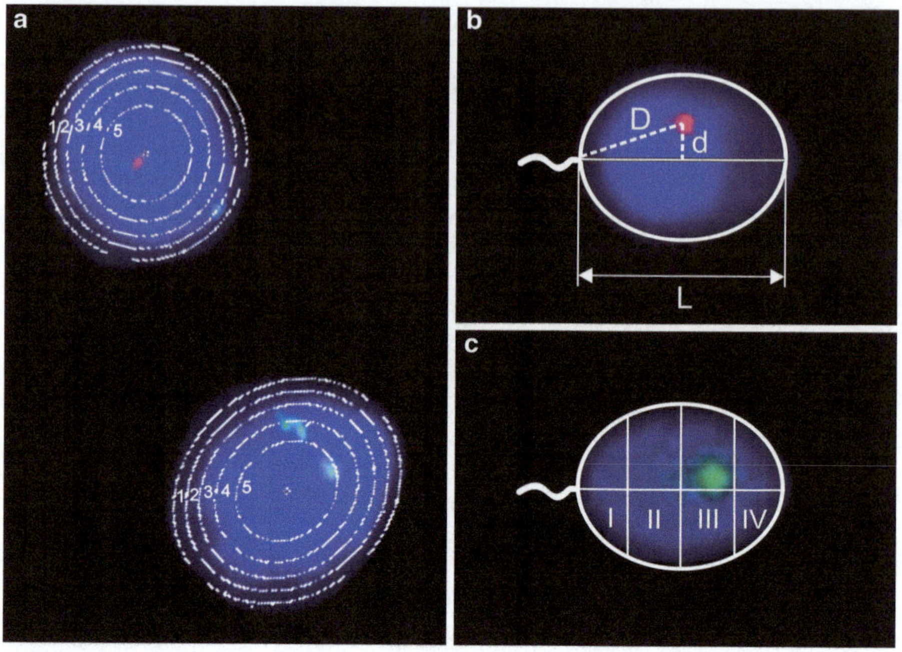

Fig. 8.1 Determination of chromosome position in human sperm nuclei using fluorescent in situ hybridization (FISH) with centromere-specific and whole chromosome painting (WCP) probes. (**a**) Centromere-specific probes with five shell template overlaid are used to determine the radial chromosome location. FISH signals are scored according to which of the five shells they appeared in; if a probe is spanned more than one shell, it is scored based on the location of the majority of its signal. Upper sperm nucleus: the chromosome 18 (*aqua*) is the outermost, the chromosome Y (*red*) is innermost; lower sperm nucleus: chromosomes X (*green*) and 18 (*aqua*) occupy intermediate positions. (From Finch et al. 2008, with kind permission from Oxford University Press.) (**b**) Centromere-specific probe (*red*) and measurements of normalized distances from FISH signal to the tail attachment point and to the long nuclear axis (D/L and d/L) are used to describe the intranuclear chromosome position. (**c**) Chromosome painting probe is used to visualize the chromosome territory (CT) of a chromosome (*green*). To determine the longitudinal position along the anterior–posterior axis (shown as a *horizontal line*), the sperm nucleus is divided into four regions, I–IV, starting from the basal side that is determined by the tail attachment site. Methods in (**b** and **c**) suggested by Zalenskaya and Zalensky (2004). Nuclear DNA (**a**–**c**) is counterstained with DAPI (*blue*)

Clustering of Centromeres into Compact Chromocenter in the Interior of the Sperm Nucleus

Immunofluorescent detection of the centromere-specific histone CENP-A and in situ hybridization with alphoid pan-centromeric DNA probes followed by confocal microscopy analysis revealed the localization of centromeres in mature human sperm nuclei. It was shown that centromeres of all nonhomologous chromosomes demonstrate a strong clustering that leads to the formation of a compact chromocenter localized to the nuclear interior (Zalensky et al. 1993, 1995). Individual centromeres within the chromocenter of the decondensed mature human sperm nucleus form different structural elements: dimers, tetramers, linear arrays, and V-shaped structures (Zalensky et al. 1993). Later on, the existence of the chromocenter in human sperm nucleus was supported (Hazzouri et al. 2000; Gurevitch et al. 2001; Finch et al. 2008; Wiland et al. 2008). Chromocenters were also found in bovine (Powell et al. 1990), rat (Meyer-Ficca et al. 1998), and mouse (Haaf and Ward 1995; Hoyer-Fender et al. 2000; Dolnik et al. 2007) sperm nuclei, suggesting similar principles of overall nuclear architecture in mammalian spermatozoa.

Peripheral Location of Telomeres in the Sperm Nucleus

The position of telomeres in human sperm cells was examined using FISH with the telomere-specific (TTAGGG)n probe. It was shown that telomeres are localized at the periphery of human sperm nuclei, and it was suggested that they form associations with the nuclear membrane (Zalensky et al. 1995). Using minimal nuclear swelling of the sperm cells pretreated with heparin in combination with DTT, it was found that in human sperm, telomeres associate into tetrameres and dimers (Zalensky et al. 1997). Telomere dimers are observed not only in humans but in five other mammals: rat, mouse, pig, horse, and cow (Zalensky et al. 1997; Meyer-Ficca et al. 1998). The progress in the studies on telomere positioning has been done after human subtelomeric chromosome- and arm-specific probes for FISH became available (Knight et al. 1997, Kingsley et al. 1997, Knight and Flint 2000). Interesting results were obtained with the use of two-color FISH with pairs of DNA probes that correspond to p- and q-arms of seven human chromosomes (metacentric and submetacentric, large and small) on sperm cells after heparin-DTT nuclear swelling (Solov'eva et al. 2004). These studies have revealed that the telomere dimer appears as the result of specific interaction of chromosome ends of the same chromosome, indicating that human sperm chromosomes are looped. It has been proposed that these interactions depend on chromosome arm-specific subtelomeric sequences and may involve protein complexes specifically binding these sequences that are not established yet.

Hairpin Configuration of Chromosomes in Human Sperm Nucleus

Using two-color FISH with microdissected probes for the whole p- and q-arms of human metacentric chromosomes 1 and 2 and submetacentric chromosome 5, the looped configuration of human sperm chromosomes proposed earlier (Zalensky et al. 1995; Solov'eva et al. 2004) was confirmed (Mudrak et al. 2005). In sperm nuclei mildly decondensed with heparin/DTT, FISH signals from p- and q-arms of the chromosomes overlapped or were located closely to each other in the antiparallel fashion inside the compact CT, while the chromosome was bent at almost 180° in the centromere region. The antiparallel position of chromosome arms produced

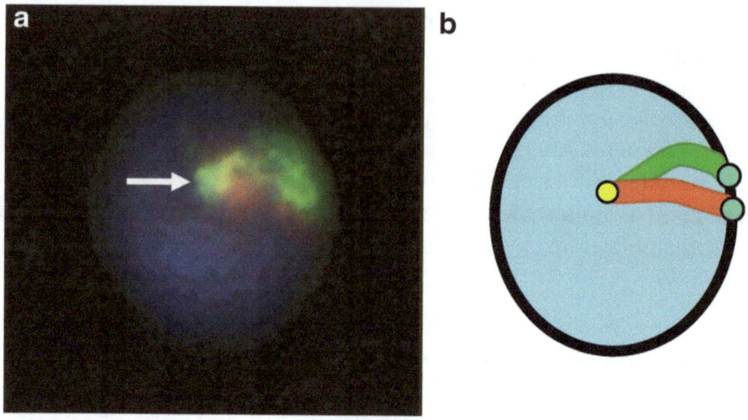

Fig. 8.2 Hairpin configuration of the chromosome in human sperm nucleus revealed by two-color FISH with arm-specific chromosome paints. (**a**) Typical pattern of hybridization: chromosome fiber is bent at 180° at the centromere (*arrow*) so that p-arm (*red*) and q-arm (*green*) of the chromosome appear aligned in the antiparallel way. This arrangement produces a looped shape of chromosomes with telomeres located close to each other. (**b**) Schematic representation of the hairpin chromosome configuration presented in (**a**). p- and q-chromosome arms are shown in *red* and *green*; the centromere, in *yellow*, and telomeres located at the nuclear periphery, in *light green*

the hairpin configuration (Fig. 8.2) of chromosomes (Mudrak et al. 2005). The hairpin shape of chromosomes was preserved in nuclei swollen to a higher degree. The hairpin conformation is most probably characteristic to the sperm chromosomes of other mammals where telomeric dimers located at the nuclear periphery were observed (Zalensky et al. 1997; Meyer-Ficca et al. 1998).

In summary, chromosome organization in human spermatozoa has the following features: (1) chromosomes are nonrandomly located in the nucleus and occupy distinct CTs; (2) centromeres of all chromosomes are joined into the chromocenter in the nuclear interior; (3) telomeres are located at the periphery of the nucleus, forming dimers and tetramers; (4) each CT appears to be stretched between its internally located centromere with p- and q-arm telomeres interacting at the periphery, so that overall chromosome configuration resembles a hairpin; (5) chromosome arms represent chromatin fibers 1,000 nm wide; each 1,000-nm fiber is composed of two rows of chromatin globules 500 nm in diameter interconnected with thinner chromatin strands (illustrated in Fig. 8.3a–c).

Chromosome Positioning in Spermatozoa and Early Embryonic Development

Complex and dynamic organization of the genome in somatic cells contributes to the regulation of nuclear processes such as DNA replication and repair, gene transcription, and RNA processing (for review, see Schneider and Grosschedl 2007).

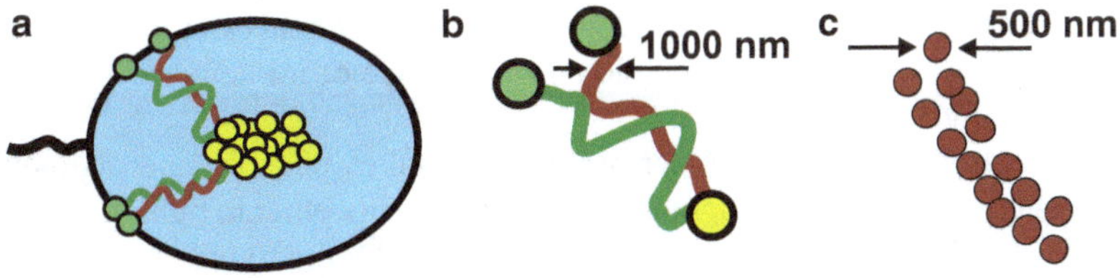

Fig. 8.3 Organization of chromosomes in human sperm nucleus. (**a**) Chromosomes (only two ones are shown) stretch between the sperm chromocenter formed by an aggregation of centromeres (*yellow circles*) in the nuclear interior and peripherally located telomeres (*light green circles*), associated into dimers. Closely located p- and q-arms produce a characteristic hairpin configuration of the sperm chromosome. (**b**) Enlarged chromosome hairpin: an individual arm presents a 1,000-nm chromatin fiber. (**c**) A 1,000-nm arm fiber consists of two rows of 500-nm chromatin globules seen in a conventional epifluorescent microscope. These globules presumably are formed by stacked nucleoprotamine toroids. Each toroid consists of a packed 50-kb DNA loop. (Overviewed in Section "DNA Packaging in Spermatozoa")

It is not entirely clear whether global genome organization in the genetically silent sperm nucleus is of any functional importance or if it arises in the process of efficient packaging of chromatin into the miniature sperm head.

Intranuclear location of CHR X was studied in spermatozoa of several mammalian species including monotremes and marsupials—nonplacental mammals evolutionarily diverged from placentals 170 and 130 million years ago, respectively. In fibrillar spermatozoa of two monotreme species—platypus and echidna—sex chromosomes are located apically near the acrosome, the place of the first contact with the egg during fertilization (Watson et al. 1996). Therefore, it was suggested that sex chromosomes were possibly the first chromosomes to be exposed to the egg cytoplasm, decondensed and remodeled by ooplasmic factors (Greaves et al. 2003). Anterior (subacrosomal) location of X chromosomes in humans (Luetjens et al. 1999; Hazzouri et al. 2000; Sbracia et al. 2002) supported this view. Although in sperm cells of two Australian marsupials, the fat-tailed dunnart and southern hairy-nosed wombat, CHR X occupied the medial position (Greaves et al. 2001, 2003), during maturation, dunnart and wombat sperm cells undergo the morphological transition to T-shape (Breed 1994), so that their medially located CHR X appears to be located in the place of spermatozoon first contact to the egg during fertilization.

The opposite point of view is that sex chromosomes are the last ones to be activated by ooplasm during fertilization (Foster et al. 2005); this is based on the analysis of sex chromosome location in spermatozoa of humans and pigs. Although the longitudinal position of CHR X in these two species (human and pig) could be defined as anterior (subacrosomal), its radial position is strictly central (Luetjens et al. 1999; Hazzouri et al. 2000; Sbracia et al. 2002; Zalenskaya and Zalensky 2004). It has been suggested that in oval-shaped porcine and human nuclei, radial positioning is more relevant, and therefore chromosomes X and Y, located deeply in the nuclear interior, are the last ones to be exposed to the ooplasm during fertilization. Similar CHR X location was observed in bovine sperm nuclei (Mudrak et al., unpublished data). Despite the diametrical points of view on timing of sex chromosome activation during fertilization, both groups of authors (Greaves et al. 2003; Foster et al. 2005) are in agreement on functional importance of intranuclear sex chromosome positioning.

Global rearrangement of nuclear architecture takes place in spermatogenesis. Individual centromeres come together to form a chromocenter in humans (Zalensky et al. 1993, 1995) and mice (Namekawa et al. 2006) by the round spermatid stage. At the same stage, repositioning of the X chromosome from the periphery to the center of the nucleus was observed also in spermatogenesis of mice (Namekawa et al. 2006) and marsupials (Namekawa et al. 2007). Similarly, in the spermatogenesis of the pig (Foster et al. 2005), relocation of CHR X from the periphery to the center occurs, when primary spermatocytes become secondary spermatocytes and then differentiate into spermatids, and this relocation seems to be important for proper sperm functioning. In sperm cell precursors such as primary spermatocytes, CHR X is located at the nuclear periphery (reviewed in Handel 2004; Foster et al. 2005).

Migration of the X chromosome may be associated with MSCI, another form of X-chromosome inactivation characteristic for male germ cells. Male MSCI is Xist independent (McCarrey et al. 2002; Turner et al. 2002), as compared to female X-inactivation, well known for placental mammals, when a noncoding RNA (Xist) decorates the entire X chromosome to initiate chromosome-wide gene silencing (Chow and Heard 2010; Lee 2010). MSCI is a special case of the more general mechanism of meiotic silencing of unsynapsed chromatin (MSUC), which silences chromosomes that fail to pair with their homologous partners and is thought to prevent the aneuploidy resulting from synaptic errors (Turner 2007).

MSCI begins during the first meiotic prophase of spermatogenesis when homologous pairing and synapsis take place. Sex chromosomes lack homology, except for a tiny pseudoautosomal region. Although autosomes undergo synapsis, sex chromosomes remain unpaired. They start condensing and eventually form a transcriptionally inactive XY-body by the mid-pachytene stage, at the nuclear periphery of primary spermatocytes (Hoyer-Fender 2003; Namekawa et al. 2006). Transcriptional silencing is achieved by a number of chromatin modifications, including histone ubiquitination, phosphorylation, methylation, and acetylation (Baarends et al. 2007). It was demonstrated that the silencing persists throughout meiosis II and spermiogenesis into mature sperm, long after dissolution of the XY-body. Inactive X and Y chromosomes, showing epigenetic similarity to the XY-body, were termed postmeiotic sex chromatin (Namekawa et al. 2006).

In postmeiotic haploid round spermatids of mice, postmeiotic sex chromatin occupies a novel compartment in the center of the nucleus, juxtaposed to the chromocenter, and adopts a distinctive chromosome configuration as compared to autosomes. Although autosomes are organized radially around the chromocenter in a Rabl configuration (Leitch 2000), the X and Y occupy compact domains that do not extend to the nuclear periphery (Namekawa et al. 2006). Similar CHR X configuration was observed in mature spermatozoa of pigs (Foster et al. 2005), bulls, and humans (Mudrak et al., unpublished data). Thus, the X-chromosome position may be critical for establishing X-chromosome inactivation.

Polarity of sperm nuclei together with the nonrandom chromosome/chromatin domain arrangement in sperm cells may lead to the sequential manner of sperm chromatin remodelling during fertilization (Foster et al. 2005). Following penetration into the oocyte, the sperm nuclear envelope starts to disperse, and sperm chromatin becomes exposed to the ooplasm. It is imaginable that chromatin bordering the disassembling nuclear envelope is decondensed and remodeled by ooplasmic factors in the first turn. Chromosomes that are closer to the dispersing nuclear envelope should be remodeled earlier than others.

During early mammalian embryogenesis, male and female genomes remain topologically separated up to the four-cell embryo stage, at the time when chromatin remodeling (Mayer et al. 2000) and programming of the appropriate patterns of parent-specific developmental gene expression occur (Fundele and Surani 1994).

Thus, the well-organized and conserved nuclear architecture in sperm may provide the mechanism for differential exposure of chromatin domains to the ooplasmic factors and the controlled activation of the male genome following fertilization.

Genome Architecture Abnormalities and Male Infertility

Male factor infertility is a heterogeneous disorder including abnormal sperm morphology, low motility and sperm count, chromosome abnormalities, deficiencies in basic chromosomal proteins, and chromatin condensation defects.

If the genome architecture is functionally important, its change is expected to be associated with an abnormal phenotype. Indeed, in somatic cells, alterations in genome architecture are correlated with epilepsy (Borden and Manuelidis 1988), some laminopathies (Misteli 2004), and cancer (Meaburn et al. 2005).

In sperm cells, deviations from normal genome architecture may affect fertilization or development. There are a few studies of genome architecture in infertile males. Finch et al. (2008) studied the radial distribution of three centromeric loci (in CHRs X, Y, and 18) in spermatozoa from infertile men diagnosed as having impaired semen parameters such as oligozoospermia, asthenozoospermia, and teratozoospermia. They found that the sex chromosome distribution pattern in these patients was altered in comparison with a control group of normozoospermic males. The distribution of X- and Y-centromeres in infertile males was close to random, whereas in the control group, centromeres were located in the center of the nucleus, which is characteristic of normal human sperm genome architecture (Zalensky et al. 1993, 1995; Zalenskaya and Zalensky 2004). The authors suggested that the distorted genome architecture in infertile males could be the consequence of impaired spermatogenic regulation. X-chromosome migration to the center of the nucleus may be related to the MSCI/MSUC process, whereas compromised semen parameters are often connected with the increase of aneuploidy level (reviewed by Tempest and Griffin 2004). Hence, Finch and coauthors speculated that the altered pattern of genome architecture in infertile males might reflect failure in sex chromosome migration from the periphery to the center and defects in MSCI/MSUC mechanisms (Finch et al. 2008).

A distorted centromere distribution pattern in spermatozoa was reported to be correlated with infertility (Olszewska et al. 2008). The authors demonstrated that although in fertile individuals the localization of centromeres (CHRs 15, 18, X, and Y) was restricted to a small area in the nuclear interior, some disturbances in the centromere area existed in sperm cells of infertile patients. In disomic sperm cells ($n+1$), centromeres of CHRs 15,15 or 18,18, or Y,Y (but not X,X) had a shifted average longitudinal position in comparison with normal sperm cells ($n=23$) (Olszewska et al. 2008).

Aberrations of genome architecture in spermatozoa do not necessarily cause infertility. A study of longitudinal and radial localization of centromeres (CHRs 7, 9, X, Y) in fertile normozoospermic males that were carriers of the reciprocal chromosome translocations t(1;7), t(7;2), t(7;13), t(7;9), t(9;14), and t(4;13) demonstrated deviations in normal chromosome positioning: the centromeres of chromosomes with translocations had a shifted intranuclear localization, which influenced the localization of other chromosomes (X and Y). The chromocenter in sperm nuclei of translocation carriers was widened toward the apical end of the nucleus in comparison with the chromocenter in control males (Wiland et al. 2008).

An increase of aneuploidy level and the degree of mosaicism in the embryo derived from parents—carriers of a balanced translocation—was reported previously (Iwarsson et al. 2000). Based on this observation, Wiland and coauthors hypothesized that distorted chromosome positioning in spermatozoa might affect the position of chromosomes during the first mitotic division of the zygote and thus could lead to aneuploidy.

Therefore, no deterministic link has been established between aberrant sperm genome architecture and male fertility. Further studies are needed to evaluate the contribution of altered sperm genome architecture to male infertility.

Sex Chromosome Positioning in Male Gametes and the Increased Rate of Sex Chromosome Aneuploidy After ICSI

Intracytoplasmic sperm injection (ICSI) is an assisted reproductive technology that is used to treat sperm-related infertility. In this technique, a single spermatozoon is injected into a mature egg, thus bypassing all the steps of normal gamete interaction. The injected spermatozoon does not undergo the acrosome reaction and preserves the intact perinuclear theca, which can cause delay in decondensation of subacrosomal chromatin.

An increased occurrence of chromosomal abnormalities in newborn children conceived through ICSI was reported (Bonduelle et al. 1998; In't Veld et al. 1995; Liebaers et al. 1995). It was suggested that one of these abnormalities, the increase in sex chromosome aneuploidy of paternal origin, might arise de novo and be a consequence of the ICSI procedure itself (Luetjens et al. 1999).

An impaired chromatin decondensation pattern was also observed in monkey (Sutovsky et al. 1996; Hewitson et al. 1999; Ramalho-Santos et al. 2000), porcine (Katayama et al. 2002), mouse (Ajduk et al. 2006), and heterologous human–hamster ICSI (Terada et al. 2000; Jones et al. 2010).

Human sex chromosomes have been shown to be preferentially located in subacrosomal areas of the sperm nucleus (Luetjens et al. 1999; Hazzouri et al. 2000; Sbracia et al. 2002; Zalenskaya and Zalensky 2004; Mudrak et al. 2012). Atypical decondensation of the subacrosomal region may lead to unbalanced remodeling of sex chromosomes (for instance, delaying their entry into S-phase) and eventually lead to errors during the first mitotic division of the zygote and chromosome loss (Terada et al. 2000).

An alternative explanation of the increased sex chromosome aneuploidy after ICSI may be an increased rate of sex chromosome aneuploidy in sperm cells. A statistically significant increase of the fraction of sperm cells with sex chromosome abnormalities was found in the semen of oligozoospermic men (Shi and Martin 2001; Sbracia et al. 2002). It was shown that normal morphology was not an absolute indicator for the selection of genetically normal sperm, and therefore the observed pregnancy failures among ICSI patients might in part be caused by the selection of aneuploid spermatozoa (Ryu et al. 2001). Prospective sperm aneuploidy testing would be beneficial for understanding the basis of chromosome anomalies and for improving the efficiency and safety of infertility therapies.

Concluding Remarks

Chromatin remodeling in the nucleus in differentiating male germ cells results in complete suppression of gene expression and supercompact packaging of the male genome into a volume of about 5 % of that of somatic cells in mature spermatozoa. Because of this, for a long time, the sperm function was considered to be limited only to the delivery of paternal genetic material to the oocyte. However, data that have emerged in the past two decades suggest that, in addition to the paternal genome, the mammalian sperm cell may bear epigenetic information important for early embryogenesis.

Human sperm chromatin adopts complex spatial organization on multiple levels starting from the packaging of DNA by protamines up to the higher-order structure of CTs. However, protamination of sperm chromatin is not complete, and it contains somatic-like domains organized in nucleosomes located at imprinted genes, developmentally transcribed genes, and genes of signaling factors (Miller et al. 2010). Similarly to chromatin in somatic cells, histone-containing domains can be marked by histone modifications (Hammoud et al. 2009; Brykczynska et al. 2010), suggesting potential paternal contribution to the epigenetic reprogramming of the zygote following fertilization.

Large-scale chromosome and nuclear architecture in sperm cells may have a functional significance as well. Individual chromosomes demonstrate nonrandom intranuclear positions, with their centromeric regions aggregated in the internal chromocenter, and specific pairing of the telomeric domains at the nuclear periphery. Such well-defined nuclear architecture can potentially meditate the order of unpacking and activation of the male genome upon fertilization, thus transmitting epigenetic information to descendant cells (Zalensky and Zalenskaya 2007).

Emerging evidence suggests that spermatozoa are not just vehicles for genome delivery, but that, similarly to somatic cells, sperm chromatin transmits epigenetic information.

However, our understanding of the functional significance of the structural organization of sperm chromatin is far from being complete. Further studies of chromatin organization in spermatozoa are essential for revealing mechanisms involved in early stages of embryogenesis.

Acknowledgments The authors thank Dr. Maria Svetlova for her valuable comments on the manuscript. This work was supported by RFBR grant 12-04-01489-a.

References

Ajduk A, Yamauchi Y, Ward MA (2006) Sperm chromatin remodeling after intracytoplasmic sperm injection differs from that of in vitro fertilization. Biol Reprod 75:442–451

Allen MJ, Bradbury EM, Balhorn R (1996) The chromatin structure of well-spread demembranated human sperm nuclei revealed by atomic force microscopy. Scanning Microsc 10:989–994

Arpanahi A, Brinkworth M, Iles D, Krawetz SA, Paradowska A, Platts AE, Saida M, Steger K, Tedder P, Miller D (2009) Endonuclease-sensitive regions of human spermatozoal chromatin are highly enriched in promoter and CTCF binding sequences. Genome Res 19:1338–1349

Baarends WM, Wassenaar E, Hoogerbrugge JW, Schoenmakers S, Sun ZW, Grootegoed JA (2007) Increased phosphorylation and dimethylation of XY body histones in the Hr6b-knockout mouse is associated with derepression of the X chromosome. J Cell Sci 120:1841–1851

Balhorn R (1982) A model for the structure of chromatin in mammalian sperm. J Cell Biol 93:298–305

Balhorn R (2007) The protamine family of sperm nuclear proteins. Genome Biol 8:227

Banerjee S, Smallwood A (1995) A chromatin model of IGF2/H19 imprinting. Nat Genet 11:237–238

Bickmore WA, Chubb JR (2003) Dispatch. Chromosome position: now, where was I? Curr Biol 13:R357–R359

Bonduelle M, Aytoz A, Van Assche E, Devroey P, Liebaers I, Van Steirteghem A (1998) Incidence of chromosomal aberrations in children born after assisted reproduction through intracytoplasmic sperm injection. Hum Reprod 13:781–782

Borden J, Manuelidis L (1988) Movement of the X chromosome in epilepsy. Science 242:1687–1691

Breed WG (1994) How does sperm meet egg?—in a marsupial. Reprod Fertil Dev 6:485–506

Brewer LR, Corzett M, Balhorn R (1999) Protamine-induced condensation and decondensation of the same DNA molecule. Science 286:120–123

Brykczynska U, Hisano M, Erkek S, Ramos L, Oakeley EJ, Roloff TC, Beisel C, Schübeler D, Stadler MB, Peters A (2010) Repressive and active histone methylation mark distinct promoters in human and mouse spermatozoa. Nat Struct Mol Biol 17:679–687

Carrell DT, Hammoud SS (2010) The human sperm epigenome and its potential role in embryonic development. Mol Hum Reprod 16:37–47

Chow JC, Heard E (2010) Nuclear organization and dosage compensation. Cold Spring Harb Perspect Biol 2:a000604

Churikov D, Zalenskaya IA, Zalensky AO (2004) Male germline-specific histones in mouse and man. Cytogenet Genome Res 105:203–214

Cremer T, Cremer C (2001) Chromosome territories, nuclear architecture and gene regulation in mammalian cells. Nat Rev Genet 2:292–301

Croft JA, Bridger JM, Boyle S, Perry P, Teague P, Bickmore WA (1999) Differences in the localization and morphology of chromosomes in the human nucleus. J Cell Biol 145:1119–1131

Dolnik AV, Pochukalina GN, Parfenov VN, Karpushev AV, Podgornaya OI, Voronin AP (2007) Dynamics of satellite binding protein CENP-B and telomere binding protein TRF2/MTBP in the nuclei of mouse spermatogenic line. Cell Biol Int 31:316–329

Finch KA, Fonseka KG, Abogrein A, Ioannou D, Handyside AH, Thornhill AR, Hickson N, Griffin DK (2008) Nuclear organization in human sperm: preliminary evidence for altered sex chromosome centromere position in infertile males. Hum Reprod 23:1263–1270

Foster HA, Abeydeera LR, Griffin DK, Bridger JM (2005) Non-random chromosome positioning in mammalian sperm nuclei, with migration of the sex chromosomes during late spermatogenesis. J Cell Sci 118:1811–1820

Fundele RH, Surani MA (1994) Experimental embryological analysis of genetic imprinting in mouse development. Dev Genet 15:515–522

Garagna S, Zuccotti M, Thornhill A, Fernandez-Donoso R, Berrios S, Capanna E, Redi CA (2001) Alteration of nuclear architecture in male germ cells of chromosomally derived subfertile mice. J Cell Sci 114:4429–4434

Gardiner-Garden M, Ballesteros M, Gordon M, Tam PP (1998) Histone- and protamine-DNA association: conservation of different patterns within the beta-globin domain in human sperm. Mol Cell Biol 18:3350–3356

Gatewood JM, Cook GR, Balhorn R, Bradbury EM, Schmid CW (1987) Sequence-specific packaging of DNA in human sperm chromatin. Science 236:962–964

Gatewood JM, Cook GR, Balhorn R, Schmid CW, Bradbury EM (1990) Isolation of four core histones from human sperm chromatin representing a minor subset of somatic histones. J Biol Chem 265:20662–20666

Geraedts JP, Pearson PL (1975) Spatial distribution of chromosomes 1 and Y in human spermatozoa. J Reprod Fertil 45:515–517

Gineitis AA, Zalenskaya IA, Yau PM, Bradbury EM, Zalensky AO (2000) Human sperm telomere-binding complex involves histone H2B and secures telomere membrane attachment. J Cell Biol 151:1591–1598

Greaves IK, Svartman M, Wakefield M, Taggart D, De Leo A, Ferguson-Smith MA, Rens W, O'Brien PC, Voullaire L, Westerman M, Graves JA (2001) Chromosomal painting detects non-random chromosome arrangement in dasyurid marsupial sperm. Chromosome Res 9:251–259

Greaves IK, Rens W, Ferguson-Smith MA, Griffin D, Marshall Graves JA (2003) Conservation of chromosome arrangement and position of the X in mammalian sperm suggests functional significance. Chromosome Res 11:503–512

Gurevitch M, Amiel A, Ben-Zion M, Fejgin M, Bartoov B (2001) Acrocentric centromere organization within the chromocenter of the human sperm nucleus. Mol Reprod Dev 60:507–516

Haaf T, Ward DC (1995) Higher order nuclear structure in mammalian sperm revealed by in situ hybridization and extended chromatin fibers. Exp Cell Res 219:604–611

Hammadeh ME, Hamad MF, Montenarh M, Fischer-Hammadeh C (2010) Protamine contents and P1/P2 ratio in human spermatozoa from smokers and non-smokers. Hum Reprod 25:2708–2720

Hammoud SS, Nix DA, Zhang H, Purwar J, Carrell DT, Cairns BR (2009) Distinctive chromatin in human sperm packages genes for embryo development. Nature (Lond) 460:473–478

Hammoud SS, Purwar J, Pflueger C, Cairns BR, Carrell DT (2010) Alterations in sperm DNA methylation patterns at imprinted loci in two classes of infertility. Fertil Steril 94:1728–1733

Handel MA (2004) The XY body: a specialized meiotic chromatin domain. Exp Cell Res 296:57–63

Hazzouri M, Rousseaux S, Mongelard F, Usson Y, Pelletier R, Faure AK, Vourc'h C, Sèle B (2000) Genome organization in the human sperm nucleus studied by FISH and confocal microscopy. Mol Reprod Dev 55:307–315

Hewitson L, Dominko T, Takahashi D, Martinovich C, Ramalho-Santos J, Sutovsky P, Fanton J, Jacob D, Monteith D, Neuringer M, Battaglia D, Simerly C, Schatten G (1999) Unique checkpoints during the first cell cycle of fertilization after intracytoplasmic sperm injection in rhesus monkeys. Nat Med 5:431–433

Hoyer-Fender S (2003) Molecular aspects of XY body formation. Cytogenet Genome Res 103:245–255

Hoyer-Fender S, Singh PB, Motzkus D (2000) The murine heterochromatin protein M31 is associated with the chromocenter in round spermatids and is a component of mature spermatozoa. Exp Cell Res 254:72–79

Hud NV, Allen MJ, Downing KH, Lee J, Balhorn R (1993) Identification of the elemental packing unit of DNA in mammalian sperm cells by atomic force microscopy. Biochem Biophys Res Commun 193:1347–1354

In't Veld P, Brandenburg H, Verhoeff A, Dhont M, Los F (1995) Sex chromosomal abnormalities and intracytoplasmic sperm injection. Lancet 346:773

Iwarsson E, Malmgren H, Inzunza J, Ahrlund-Richter L, Sjoblom P, Rosenlund B, Fridtrom M, Hovatta O, Nordenskjold M, Blennow E (2000) Highly abnormal cleavage divisions in preimplantation embryo from translocations carriers. Prenat Diagn 20:1038–1047

Johnson GD, Lalancette C, Linnemann AK, Leduc F, Boissonneault G, Krawetz SA (2011) The sperm nucleus: chromatin, RNA, and the nuclear matrix. Reproduction 141:21–36

Jones EL, Mudrak O, Zalensky AO (2010) Kinetics of human male pronuclear development in a heterologous ICSI model. J Assist Reprod Genet 27:277–283

Joshi NV, Medina H, Colasante C, Osuna A (2000) Ultrastructural investigation of human sperm using atomic force microscopy. Arch Androl 44:51–57

Katayama M, Koshida M, Miyake M (2002) Fate of the acrosome in ooplasm in pigs after IVF and ICSI. Hum Reprod 17:2657–2664

Kingsley K, Wirth J, van der Maarel S, Freier S, Ropers HH, Haaf T (1997) Complex FISH probes for the subtelomeric regions of all human chromosomes: comparative hybridization of CEPH YACs to chromosomes of the Old World monkey *Presbytis cristata* and great apes. Cytogenet Cell Genet 78:12–19

Knight SJ, Flint J (2000) Perfect endings: a review of subtelomeric probes and their use in clinical diagnosis. J Med Genet 37:401–409

Knight SJ, Horsley SW, Regan R, Lawrie NM, Maher EJ, Cardy DL, Flint J, Kearney L (1997) Development and clinical application of an innovative fluorescence in situ hybridization technique which detects submicroscopic rearrangements involving telomeres. Eur J Hum Genet 5:1–8

Lanctôt C, Cheutin T, Cremer M, Cavalli G, Cremer T (2007) Dynamic genome architecture in the nuclear space: regulation of gene expression in three dimensions. Nat Rev Genet 8:104–115

Lee JT (2010) The X as model for RNA's niche in epigenomic regulation. Cold Spring Harb Perspect Biol 2:a003749

Leitch AR (2000) Higher levels of organization in the interphase nucleus of cycling and differentiated cells. Microbiol Mol Biol Rev 64:138–152

Li Z, Yang J, Huang H (2006) Oxidative stress induces H2AX phosphorylation in human spermatozoa. FEBS Lett 580:6161–6168

Liebaers I, Bonduelle M, Van Assche E, Devroey P, Van Steirteghem A (1995) Sex chromosome abnormalities after intracytoplasmic sperm injection. Lancet 346:1095

Luetjens CM, Payne C, Schatten G (1999) Non-random chromosome positioning in human sperm and sex chromosome anomalies following intracytoplasmic sperm injection. Lancet 353:1240

Mahadevaiah SK, Turner JM, Baudat F, Rogakou EP, de Boer P, Blanco-Rodriguez J, Jasin M, Keeney S, Bonner WM, Burgoyne PS (2001) Recombinational DNA double-strand breaks in mice precede synapsis. Nat Genet 27:271–276

Manvelyan M, Hunstig F, Bhatt S, Mrasek K, Pellestor F, Weise A, Simonyan I, Aroutiounian R, Liehr T (2008) Chromosome distribution in human sperm—a 3D multicolor banding-study. Mol Cytogenet 1:25

Mayer W, Smith A, Fundele R, Haaf T (2000) Spatial separation of parental genomes in preimplantation mouse embryos. J Cell Biol 148:629–634

McCarrey JR, Watson C, Atencio J, Ostermeier GC, Marahrens Y, Jaenisch R, Krawetz SA (2002) X-chromosome inactivation during spermatogenesis is regulated by an Xist/Tsix-independent mechanism in the mouse. Genesis 34:257–266

Meaburn KJ, Levy N, Toniolo D, Bridger JM (2005) Chromosome positioning is largely unaffected in lymphoblastoid cell lines containing emerin or A-type lamin mutations. Biochem Soc Trans 33:1438–1440

Meister P, Mango SE, Gasser SM (2011) Locking the genome: nuclear organization and cell fate. Curr Opin Genet Dev 21:167–174

Meistrich ML, Mohapatra B, Shirley CR, Zhao M (2003) Roles of transition nuclear proteins in spermiogenesis. Chromosoma (Berl) 111:483–488

Meyer-Ficca M, Muller-Navia J, Scherthan H (1998) Clustering of pericentromeres initiates in step 9 of spermiogenesis of the rat (*Rattus norvegicus*) and contributes to a well defined genome architecture in the sperm nucleus. J Cell Sci 111:1363–1370

Miller D, Brinkworth M, Iles D (2010) Paternal DNA packaging in spermatozoa: more than the sum of its parts? DNA, histones, protamines and epigenetics. Reproduction 139:287–301

Misteli T (2004) Spatial positioning: a new dimension in genome function. Cell 119:153–156

Misteli T (2007) Beyond the sequence: cellular organization of genome function. Cell 128:787–800

Mudrak OS, Nazarov IB, Jones EL, Zalensky AO (2012) Positioning of chromosomes in human spermatozoa is determined by ordered centromere arrangement. PLoS ONE 7: e52944

Mudrak O, Tomilin N, Zalensky A (2005) Chromosome architecture in the decondensing human sperm nucleus. J Cell Sci 118:4541–4550

Mudrak O, Zalenskaya I, Zalensky O (2011) Organization of chromosomes during spermatogenesis and in mature sperm. In: Rousseaux S, Khochbin S (eds) Epigenetics and human reproduction, 1st edn. Springer, Berlin

Nagele RG, Freeman T, McMorrow L, Thomson Z, Kitson-Wind K, Lee H (1999) Chromosomes exhibit preferential positioning in nuclei of quiescent human cells. J Cell Sci 112:525–535

Namekawa SH, Park PJ, Zhang LF, Shima JE, McCarrey JR, Griswold MD, Lee JT (2006) Postmeiotic sex chromatin in the male germline of mice. Curr Biol 16:660–667

Namekawa SH, VandeBerg JL, McCarrey JR, Lee JT (2007) Sex chromosome silencing in the marsupial male germ line. Proc Natl Acad Sci USA 104:9730–9735

Nanassy L, Liu L, Griffin J, Carrell DT (2011) The clinical utility of the protamine 1/protamine 2 ratio in sperm. Protein Pept Lett 18:772–777

Nazarov IB, Shlyakhtenko LS, Lyubchenko YL, Zalenskaya IA, Zalensky AO (2008) Sperm chromatin released by nucleases. Syst Biol Reprod Med 54:37–46

Olszewska M, Wiland E, Kurpisz M (2008) Positioning of chromosome 15, 18, X and Y centromeres in sperm cells of fertile individuals and infertile patients with increased level of aneuploidy. Chromosome Res 16:875–890

Pai DA, Engelke DR (2010) Spatial organization of genes as a component of regulated expression. Chromosoma (Berl) 119:13–25

Parada LA, Sotiriou S, Misteli T (2004) Spatial genome organization. Exp Cell Res 296:64–70

Pittoggi C, Renzi L, Zaccagnini G, Cimini D, Degrassi F, Giordano R, Magnano AR, Lorenzini R, Lavia P, Spadafora C (1999) A fraction of mouse sperm chromatin is organized in nucleosomal hypersensitive domains enriched in retroposon DNA. J Cell Sci 112:3537–3548

Powell D, Cran DG, Jennings C, Jones R (1990) Spatial organization of repetitive DNA sequences 500 in the bovine sperm nucleus. J Cell Sci 97:185–191

Ramalho-Santos J, Sutovsky P, Simerly C, Oko R, Wessel GM, Hewitson L, Schatten G (2000) ICSI choreography: fate of sperm structures after monospermic rhesus ICSI and first cell cycle implications. Hum Reprod 15:2610–2620

Ryu HM, Lin WW, Lamb DJ, Chuang W, Lipshultz LI, Bischoff FZ (2001) Increased chromosome X, Y, and 18 nondisjunction in sperm from infertile patients that were identified as normal by strict morphology: implication for intracytoplasmic sperm injection. Fertil Steril 76:879–883

Sbracia M, Baldi M, Cao D, Sandrelli A, Chiandetti A, Poverini R, Aragona C (2002) Preferential location of sex chromosomes, their aneuploidy in human sperm, and their role in determining sex chromosome aneuploidy in embryos after ICSI. Hum Reprod 17:320–324

Schneider R, Grosschedl R (2007) Dynamics and interplay of nuclear architecture, genome organization, and gene expression. Genes Dev 21:3027–3043

Shi Q, Martin RH (2001) Aneuploidy in human spermatozoa: FISH analysis in men with constitutional chromosomal abnormalities, and in infertile men. Reproduction 121:655–666

Simon L, Castillo J, Oliva R, Lewis SE (2011) Relationships between human sperm protamines, DNA damage and assisted reproduction outcomes. Reprod Biomed Online 23:724–734

Singleton S, Mudrak O, Morshedi M, Oehninger S, Zalenskaya I, Zalensky A (2007) Characterisation of a human sperm cell subpopulation marked by the presence of the TSH2B histone. Reprod Fertil Dev 19:392–397

Solov'eva L, Svetlova M, Bodinski D, Zalensky AO (2004) Nature of telomere dimers and chromosome looping in human spermatozoa. Chromosome Res 12:817–823

Sotolongo B, Huang TT, Isenberger E, Ward WS (2005) An endogenous nuclease in hamster, mouse, and human spermatozoa cleaves DNA into loop-sized fragments. J Androl 26:272–280

Sun HB, Yokota H (1999) Correlated positioning of homologous chromosomes in daughter fibroblast cells. Chromosome Res 7:603–610

Sutovsky P, Hewitson L, Simerly CR, Tengowski MW, Navara CS, Haavisto A, Schatten G (1996) Intracytoplasmic sperm injection for *Rhesus* monkey fertilization results in unusual chromatin, cytoskeletal, and membrane events, but eventually leads to pronuclear development and sperm aster assembly. Hum Reprod 11:1703–1712

Tempest HG, Griffin DK (2004) The relationship between male infertility and increased levels of sperm disomy. Cytogenet Genome Res 107:83–94

Terada Y, Luetjens CM, Sutovsky P, Schatten G (2000) Atypical decondensation of the sperm nucleus, delayed replication of the male genome, and sex chromosome positioning following intracytoplasmic human sperm injection (ICSI) into golden hamster eggs: does ICSI itself introduce chromosomal anomalies? Fertil Steril 74:454–460

Tilgen N, Guttenbach M, Schmid M (2001) Heterochromatin is not an adequate explanation for close proximity of interphase chromosomes 1–Y, 9–Y, and 16–Y in human spermatozoa. Exp Cell Res 265:283–287

Tsend-Ayush E, Dodge N, Mohr J, Casey A, Himmelbauer H, Kremitzki CL, Schatzkamer K, Graves T, Warren WC, Grützner F (2009) Higher-order genome organization in platypus and chicken sperm and repositioning of sex chromosomes during mammalian evolution. Chromosoma (Berl) 118:53–69

Tunc O, Thompson J, Tremellen K (2009) Improvement in sperm DNA quality using an oral antioxidant therapy. Reprod Biomed Online 18:761–768

Turner JM (2007) Meiotic sex chromosome inactivation. Development (Camb) 134:1823–1831

Turner JM, Mahadevaiah SK, Elliott DJ, Garchon HJ, Pehrson JR, Jaenisch R, Burgoyne PS (2002) Meiotic sex chromosome inactivation in male mice with targeted disruptions of Xist. J Cell Sci 115:4097–4105

van der Heijden GW, Ramos L, Baart EB, van den Berg IM, Derijck AA, van der Vlag J, Martini E, de Boer P (2008) Sperm-derived histones contribute to zygotic chromatin in humans. BMC Dev Biol 8:34

Ward WS (2010) Function of sperm chromatin structural elements in fertilization and development. Mol Hum Reprod 16:30–36

Ward MA, Ward WS (2004) A model for the function of sperm DNA degradation. Reprod Fertil Dev 16:547–554

Watson JM, Meyne J, Graves JA (1996) Ordered tandem arrangement of chromosomes in the sperm heads of monotreme mammals. Proc Natl Acad Sci USA 93:10200–10205

Wiland E, Zegalo M, Kurpisz M (2008) Interindividual differences and alterations in the topology of chromosomes in human sperm nuclei of fertile donors and carriers of reciprocal translocations. Chromosome Res 16:291–305

Wykes SM, Krawetz SA (2003) The structural organization of sperm chromatin. J Biol Chem 278:29471–29477

Zalenskaya IA, Zalensky AO (2004) Non-random positioning of chromosomes in human sperm nuclei. Chromosome Res 12:163–173

Zalenskaya IA, Bradbury EM, Zalensky AO (2000) Chromatin structure of telomere domain in human sperm. Biochem Biophys Res Commun 279:213–218

Zalensky A, Zalenskaya I (2007) Organization of chromosomes in spermatozoa: an additional layer of epigenetic information? Biochem Soc Trans 35:609–611

Zalensky AO, Breneman JW, Zalenskaya IA, Brinkley BR, Bradbury EM (1993) Organization of centromeres in the decondensed nuclei of mature human sperm. Chromosoma (Berl) 102:509–518

Zalensky AO, Allen MJ, Kobayashi A, Zalenskaya IA, Balhorn R, Bradbury EM (1995) Well-defined genome architecture in the human sperm nucleus. Chromosoma (Berl) 103:577–590

Zalensky AO, Tomilin NV, Zalenskaya IA, Teplitz RL, Bradbury EM (1997) Telomere–telomere interactions and candidate telomere binding protein(s) in mammalian sperm cells. Exp Cell Res 232:29–41

Zalensky AO, Siino JS, Gineitis AA, Zalenskaya IA, Tomilin NV, Yau P, Bradbury EM (2002) Human testis/sperm-specific histone H2B (hTSH2B). Molecular cloning and characterization. J Biol Chem 277:43474–43480

Chapter 9
Interphase Chromosome-Specific Multicolor Banding

Ivan Y. Iourov, Thomas Liehr, Svetlana G. Vorsanova, and Yuri B. Yurov

Abstract Interphase chromosome-specific multicolor banding (ICS-MCB) is a FISH-based technique for studying interphase chromosomes in their integrity at molecular resolution in single cells. To date, this is the only method providing for visualizing the whole chromosome and its specific regions together at any stage of the cell cycle. Offering such opportunity, ICS-MCB has repeatedly proven effective in molecular cytogenetic studies of interphase chromosomes (i.e., surveying genomic variations at the chromosomal level in different human tissues and characterizing interphase genome/chromosome organization and behavior). The intention of this chapter is to describe the basics of ICS-MCB procedure and its applications in different biomedical fields. The advantages of this technique allow speculations that it has to become a method of choice in somatic cell genetics as well as in cell and chromosome biology.

I.Y. Iourov (✉)
Mental Health Research Center, Russian Academy of Medical Sciences, Moscow 119152, Russia

Institute of Pediatrics and Children Surgery, Ministry of Health, Moscow, Russia
e-mail: ivan.iourov@gmail.com

T. Liehr
Jena University Hospital, Institute of Human Genetics, Friedrich Schiller University, Jena, Germany

S.G. Vorsanova • Y.B. Yurov
Mental Health Research Center, Russian Academy of Medical Sciences, Moscow 119152, Russia

Institute of Pediatrics and Children Surgery, Ministry of Health, Moscow, Russia

Moscow City University of Psychology and Education, Moscow, Russia

Y.B. Yurov et al. (eds.), *Human Interphase Chromosomes: Biomedical Aspects*, DOI 10.1007/978-1-4614-6558-4_9, © Springer Science+Business Media, LLC 2013

Introduction

Technical advances in molecular cytogenetics were the starting point of accumulating empirical data on interphase chromosomes. Based on fluorescence in situ hybridization (FISH), molecular cytogenetic techniques are successfully used for visualizing specific genomic loci (molecular cytogenetic diagnosis) and chromosome territories in interphase nuclei (Iourov et al. 2009c; Liehr 2009; Cremer and Cremer 2010; Vorsanova et al. 2010a, b). However, the evolving understanding of the role played by interphase chromosome plasticity in genome behavior (Claussen 2005) and the extensive variability of the cellular genome (Iourov et al. 2008a, b, 2010; Yurov et al. 2010) has forced the search for new technological solutions for a higher-resolution interphase molecular cytogenetic analysis. To achieve such resolution, a technique granting simultaneous visualization of the whole chromosome and its regions in a given nucleus (i.e., banding of interphase chromosomes in single cells) has been required. Actually, interphase chromosome-specific multicolor banding (ICS-MCB) has become the solution providing for studying interphase chromosomes in their integrity at molecular resolution (Iourov et al. 2006a, 2007; Iourov 2012). The technique has been further found to be applicable in studying somatic genome variations manifesting as chromosomal gain/loss (aneuploidy) and genome/chromosome instability (Yurov et al. 2007, 2008; Iourov et al. 2009a, b), as well as in analysis of interphase chromosome (genome) organization (Manvelyan et al. 2008a, b, 2009).

Using ICS-MCB it is possible to determine chromosome numbers, structure, and positioning in interphase nuclei. Specifically, one is able to determine structural variations and the arrangement of differentially painted chromosomal regions. The initial success of the method encouraged us to propose previously that ICS-MCB is a tool of choice in attempts to determine structural and functional chromosome/genome variations in interphase (Iourov et al. 2007, 2009c; Liehr 2009; Liehr et al. 2010; Vorsanova et al. 2010a; Iourov 2012). Notwithstanding, some questions still arise regarding the possibilities of this technique that hinder its wide use in biomedical research. Consequently, we provide here a thorough description of ICS-MCB with special attention paid to the advantages, disadvantages, and areas of established and potential applications.

The Basics of the Procedure

Multicolor banding (MCB) is a FISH-based approach toward banding chromosomes, which is defined as a molecular cytogenetic technique depicting simultaneously several chromosomal regions (or subregions) smaller than a chromosome arm (Liehr et al. 2010). MCB was developed through utilizing microdissected DNA probes and was initially applied to metaphase chromosomes for studying genome variations at the chromosomal/subchromosomal level (Liehr et al. 2002). Because MCB produces DNA-specific banding, it possesses a higher resolution as to cytogenetic banding (i.e., G-banding) and provides for identifying the DNA-based chromosomal structure (Liehr et al. 2002, 2010). MCB adaptation to studying numbers and arrangement of interphase chromosomes (ICS-MCB elaboration) has become possible because of original imaging approaches allowing us to determine chromosomal axis, region (locus) location, and arrangement in differently shaped interphase nuclei (Iourov et al. 2006a, 2007; Manvelyan et al. 2008a). In general, apart from imaging, ICS-MCB resembles metaphase MCB, but a specificity of ICS-MCB procedure does exist.

Both MCB and ICS-MCB are based on changing fluorescence intensity ratios along the chromosome axis. By means of corresponding software or an "in-house" analysis, different pseudocolors are assigned to specific chromosomal regions. The latter makes available differentiating between chromosomal regions, which are referred to specifically painted areas at the band or subband level. The banding becomes visible because three to five differently labeled microdissected probes are used, which are hybridizing along the chromosomal axis and overlap with each other. The color sequence results from painting of chromosomal regions by these probes per se and overlappings of probe signals, which are assigned to pseudocolors according to the fluorescence ratios within the area of signal overlapping. Although the MCB concept suggests having specific software for successful application, ICS-MCB results can be alternatively acquired through scoring signals by simple visual analysis, especially during evaluation of chromosome numbers in large cell populations (Liehr et al. 2002, 2010; Yurov et al. 2007; Liehr 2009; Iourov et al. 2006a, c, 2007, 2009c; Vorsanova et al. 2010a). For an extended reference list, one can refer to http://www.fish.uniklinikum-jena.de/mFISH/banding/MCB.html.

The resolution of interphase molecular cytogenetic analysis can be affected by chromosome spatial organization (Vorsanova et al. 2010a). Chromosome associations (somatic homologous chromosome pairing), nucleus flattening during cell suspension preparation, and chromosomal DNA diffusion within the nuclear milieu all cause misinterpretation of interphase FISH results (Iourov et al. 2006b, 2009c). To solve this problem by ICS-MCB, imaging processing is required. First, image pre-processing for quantification of relative intensities is performed. Second, digital image analysis by quantification is made. These procedures allow (1) determining the chromosomal axis by removing background from chromosomal DNA diffusion, (2) defining the location and volume occupied by each labeled chromosomal region, and (3) differentiating between specific arrangement of chromosomal loci in interphase (inside the nuclear volume) and chromosome abnormalities (Iourov et al. 2007). Alternatively, software for digital 3D analysis of microscopically visible structures offers options similar to (1) and (2), which are required for studies of nuclear chromosome organization via ICS-MCB (Manvelyan et al. 2008a, b, 2009). Figure 9.1 demonstrates ICS-MCB on human interphase nuclei and the way to increase efficiency and resolution by using different quantification protocols (Iourov et al. 2007).

To consider established and potential applications of ICS-MCB, advantages and disadvantages of this method should be listed. Classical (metaphase) MCB provides

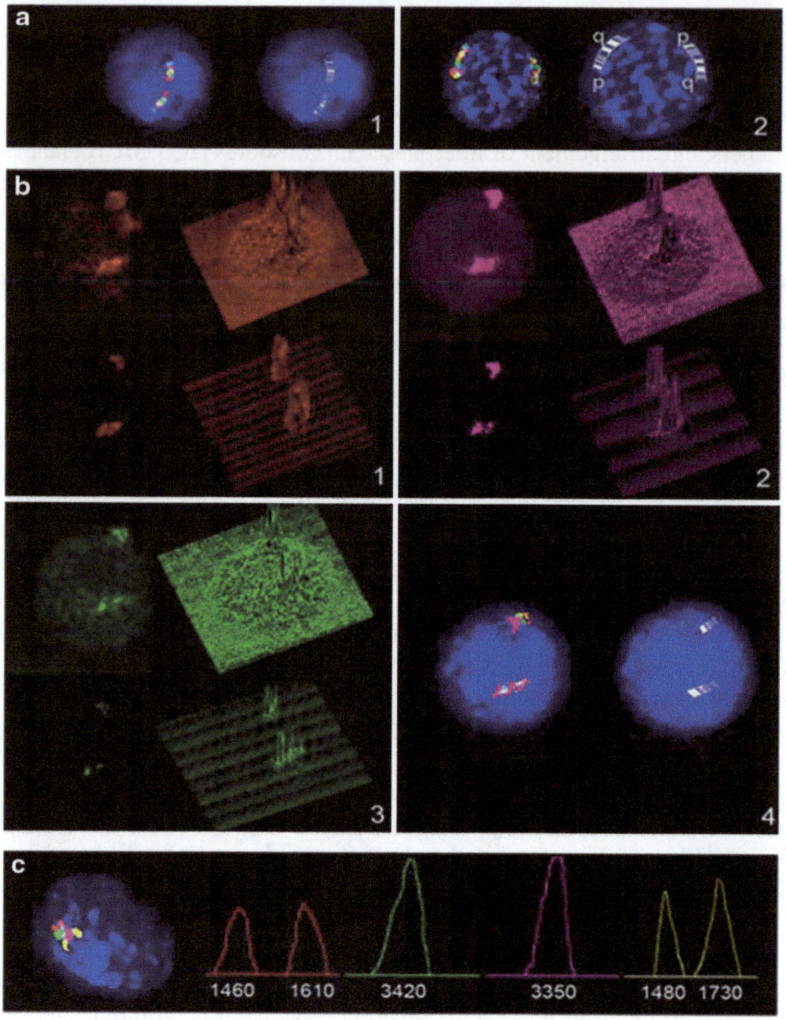

Fig. 9.1 Interphase chromosome-specific multicolor banding (ICS-MCB) on human interphase nuclei. (**a**) (*1*) ICS-MCB probe set for chromosome 14 characterizing the localization and orientation of the two homologues (*right*: pseudocolor results; *left*: schematically depicted results as G-banding ideograms); (*2*) corresponding results obtained with an ICS-MCB with probe set for chromosome 7. (**b**) ICS-MCB probe-set for chromosome 21 depicted as 3D-intensity profiles (*parts 1–3*) and as depicted in *part 4*. 3D-intensity profiles can be used to define chromosome integrity. The original results can be evaluated by equalizing the background within nuclear area to zero: SpectrumOrange signals specific for 21q11.1-q21 (*1*), TexasRed signals specific for 21q21-qter (*2*), and SpectrumGreen signals located in 21q21-q22 (*3*) were treated in this manner. Original results are shown in the upper an. equalized ones in the lower lines, each. In (*4*), results corresponding to these are shown in a figure obtained with an ICS-MCB with probe set for chromosome 21. (**c**) Results of ICS-MCB obtained with a probe set for chromosome 16 in pseudocolor depiction (*left*) together with the corresponding quantitative fluorescence in situ hybridization (QFISH) results (*right*). For the latter, the relative intensity values for each color channel used (SpectrumOrange, SpectrumGreen, TexasRed, and Cyanine5) are given in fluorescence intensity curves. According to pseudocolor depiction, the presence of one or two chromosomes 16 on this nucleus could have been suggested. QFISH clearly shows that two overlapping chromosomes are present, because the intensities of two discrete signals of SpectrumOrange and Cyanine5 are approximately two times lower than those of single signals in SpectrumGreen and TexasRed. (From Iourov et al. 2007. Reproduced with permission of Elsevier B.V. in the format reuse in a book/textbook via Copyright Clearance Center)

for analyzing chromosomes irrespective of their condensation grades (Liehr et al. 2002). It is also applicable for ICS-MCB, which is able to depict a homologous chromosome pair at any cell-cycle stage in any cell type regardless of cellular size, chromatin compaction level, density of chromosome positioning, etc. (Iourov et al. 2006a, 2009a, b; Manvelyan et al. 2008a, b, 2009; Liehr 2009). In contrast to other types of interphase FISH assays, ICS-MCB shows not only specific genomic loci or ambiguous chromosome territories (Vorsanova et al. 2010a), but a pair of discrete homologous chromosomes "band by band." The possibility to visualize simultaneously the entire chromosome, its regions (size, ~3–15 Mb), and intranuclear arrangement defines ICS-MCB as the unique method for studying chromosome abnormalities and nuclear organization in all cell types, at different cell-cycle stages, and at molecular resolutions. Because DNA–DNA FISH underlies ICS-MCB, the technique is able to provide qualitative and quantitative definition of chromosomal DNA-based structure important in evaluations of nuclear architecture and discriminations between structural alterations to chromosomes and chromosomal spatial positioning (Lemke et al. 2002; Weise et al. 2002; Iourov et al. 2006c, 2009c). ICS-MCB disadvantages are referred to the impossibility of analyzing more than one homologous chromosome pair per assay and problems with initial interpretation of digital microscopic images. However, these interpretation problems are relatively easily solved using quantification or additional software for three-dimensional (3D) microscopic analysis (Iourov et al. 2007; Manvelyan et al. 2008a).

Applications

According to our experience, there are two essential directions for ICS-MCB application: identification of chromosome abnormalities or genomic variations at the chromosomal or subchromosomal level at the single-cell level (Iourov et al. 2006a, c, 2008a, b, 2009a, b; Vorsanova et al. 2010b; Yurov et al. 2007, 2008) and definition of DNA-based structure of human chromosomes in interphase or analysis of nuclear chromosome organization (Lemke et al. 2002; Weise et al. 2002; Iourov et al. 2007; Manvelyan et al. 2008a, b, 2009; Liehr 2009). The former direction is relevant to basic and applied research targeted at defining the role of somatic genome variations in health and disease (Iourov et al. 2006c, 2008a, b, 2010; Yurov et al. 2010) and developing approaches to molecular cytogenetic diagnosis of chromosomal mosaicism and instability in interphase (Vorsanova et al. 2010b), whereas the latter is an important issue in basic biomedical research aimed at uncovering the meaning of spatial intranuclear genome organization at the chromosomal level (Claussen 2005; Cremer and Cremer 2010).

The cellular genome is highly variable and can undergo ontogenetic, clonal, and sporadic changes, but the intrinsic causes and consequences of this phenomenon are hardly known (Iourov et al. 2010; Yurov et al. 2010). To obtain further insights into somatic cell genetics and genomics, it is necessary to determine the amount of cells possessing the deviant genome in different cell types of diseased and healthy tissues.

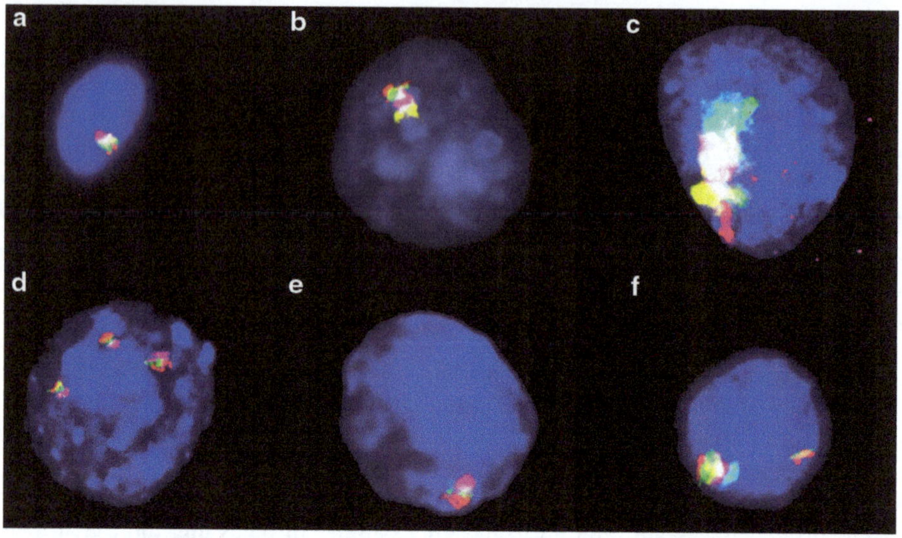

Fig. 9.2 Molecular cytogenetic analyses of the developing and adult human brain by ICS-MCB. (**a**) Loss of chromosome 18 (monosomy) in a cell isolated from telencephalic regions of the fetal brain. (**b**) Loss of chromosome 16 (monosomy) in a cell isolated from the cerebral cortex of the normal human brain (**c**) Loss of chromosome 1 (monosomy) in a cell isolated from the cerebral cortex of the schizophrenia brain. (**d**) Gain of chromosome 21 (trisomy) in a cell isolated from the cerebral cortex of an Alzheimer's disease brain. (**e**) Loss of chromosome 21 (monosomy) in a cell isolated from the cerebellum of the ataxia-telangiectasia brain. (**f**) Chromosome instability in the cerebellum of the ataxia-telangiectasia brain manifesting as the presence of a rearranged chromosome 14 or der(14)(14pter->14q12:)

Once acquired, the data would be useful for molecular cytogenetic diagnosis of somatic genome variations as a catalogue of disease-causing and benign intercellular genomic variations as a platform for developing techniques for studying cellular genome. In this context, ICS-MCB is so far the only possibility to determine single-cell genomic variations at chromosomal and subchromosomal levels (Iourov 2012). Thus, ICS-MCB can be used for molecular cytogenetic diagnosis of somatic chromosomal mosaicism in uncultured tissues (detection rate, <0.1 % or one interphase nucleus in cases of pseudomosaicism) (Iourov et al. 2008a, 2009c; Vorsanova et al. 2010b), determining pathogenetic mechanisms of a wide spectrum of human diseases mediated by somatic genome variations (for review, see Iourov et al. 2008a, 2010; Yurov et al. 2010), and identification of chromosome instability in interphase cells [chromosome instability manifested as aneuploidy/polyploidy, structural chromosome abnormalities, alterations to chromosome morphology (fragile sites, interphase chromosome breakage), etc.] (Iourov et al. 2009a, b). Figure 9.2 shows original examples of ICS-MCB applications for studying chromosome number variations (i.e., aneuploidy) and instability (structural rearrangements caused by interphase chromosome breakage). In total, ICS-MCB appears to have practical implications for molecular cytogenetic diagnosis of chromosome abnormalities (or chromosomal mosaicism) in medical genetics/cytogenetics, including oncocytogenetics and reproductive cytogenetics. Furthermore, studying intercellular genome variations is strongly required in such biomedical areas as ontogenetic (aging and development) and cancer research, single-cell biology (genomics), neuroscience, and personalized medicine.

Recent success in studying interphase genome organization by ICS-MCB has made possible speculations about a higher resolution of this technique for determining spatial arrangement of chromosomes inside the nucleus compared to other interphase FISH assays (Manvelyan et al. 2008a, b, 2009). As noted before, simultaneous analysis of the whole chromosome and its regions is the unique feature of ICS-MCB. Providing for DNA-based interphase chromosome structure with molecular resolution (Lemke et al. 2002; Weise et al. 2002), ICS-MCB is likely to be the essential technique for evaluating positioning of specific chromosomal regions in relationship to the whole chromosome and different nuclear compartments. This opportunity seems to be required for acquiring more profound knowledge concerning the real meaning of spatial chromosome positioning in interphase. Moreover, the visualization of chromosomal associations, which are likely to be involved in epigenetic genome regulation and susceptibility to somatic chromosomal rearrangements (Cremer and Cremer 2010), is much more efficiently performed by ICS-MCB (Iourov et al. 2006a, 2007; Fig. 9.1) compared to techniques showing ambiguous chromosome territories (chromosomal regions are not visible) or ligation-based chromosome conformation capture (number of chromosomal loci is limited and only small genomic fragments, sized as an average PCR product, can be evaluated). Therefore, one can further speculate that studies in chromosome and cell biology as well as somatic genetics, which are dedicated to analysis of chromosome structural and functional organization at the single-cell level, can gain much from ICS-MCB application.

Conclusion

ICS-MCB offers an opportunity to visualize together individual interphase chromosomes and their regions. The advantages of this technique (i.e., resolution, ability to study interphase chromosomes at any cell-cycle stage in any cell type) make it highly applicable in chromosome and cell biology. The method may be used as for chromosomal analysis on large cell populations, such as for definition of DNA-based structure of chromosomal loci in single interphase nuclei. Using ICS-MCB, one can acquire new data on somatic genome variations at the chromosomal level in any cell type (including structural chromosome abnormalities or instabilities, which are rarely detected by single-cell analyses of uncultured cells) and spatial chromosome/genome nuclear organization. Taking into account the data already obtained, ICS-MCB may be considered a kind of technological achievement in chromosome biology. To this end, the technique described here deserves to become a method of choice in genetics, cell, genome, and chromosome biology.

Acknowledgements The authors are supported by DLR/BMBF (RUS 2011-2013) and RFBR grant 12-04-00215-a (Russian Federation, 2012–2014). The authors also acknowledge the NICHD Brain and Tissue Bank for Developmental Disorders at the University of Maryland, Baltimore, MD, USA, for providing the brain tissue samples studied by ICS-MCB (Fig. 9.2). Dr. IY Iourov is supported by the Grant of the President of the Russian Federation MD-4401.2013.7.

References

Claussen U (2005) Chromosomics. Cytogenet Genome Res 111:101–106

Cremer T, Cremer M (2010) Chromosome territories. Cold Spring Harb Perspect Biol 2:a003889

Iourov IY (2012) To see an interphase chromosome or: how a disease can be associated with specific nuclear genome organization. Biodiscovery 4:5. doi:10.7750/BioDiscovery.2012.4.5

Iourov IY, Liehr T, Vorsanova SG, Kolotii AD, Yurov YB (2006a) Visualization of interphase chromosomes in postmitotic cells of the human brain by multicolour banding (MCB). Chromosome Res 14:223–229

Iourov IY, Vorsanova SG, Pellestor F, Yurov YB (2006b) Brain tissue preparations for chromosomal PRINS labeling. Methods Mol Biol 334:123–132

Iourov IY, Vorsanova SG, Yurov YB (2006c) Chromosomal variation in mammalian neuronal cells: known facts and attractive hypotheses. Int Rev Cytol 249:143–191

Iourov IY, Liehr T, Vorsanova SG, Yurov YB (2007) Interphase chromosome-specific multicolor banding (ICS-MCB): a new tool for analysis of interphase chromosomes in their integrity. Biomol Eng 24:415–417

Iourov IY, Vorsanova SG, Yurov YB (2008a) Chromosomal mosaicism goes global. Mol Cytogenet 1:26

Iourov IY, Vorsanova SG, Yurov YB (2008b) Molecular cytogenetics and cytogenomics of brain diseases. Curr Genomics 9:452–465

Iourov IY, Vorsanova SG, Liehr T, Kolotii AD, Yurov YB (2009a) Increased chromosome instability dramatically disrupts neural genome integrity and mediates cerebellar degeneration in the ataxia-telangiectasia brain. Hum Mol Genet 18:2656–2669

Iourov IY, Vorsanova SG, Liehr T, Yurov YB (2009b) Aneuploidy in the normal, Alzheimer's disease and ataxia-telangiectasia brain: differential expression and pathological meaning. Neurobiol Dis 34:212–220

Iourov IY, Vorsanova SG, Soloviev IV, Yurov YB (2009c) Interphase FISH: detection of intercellular genomic variations and somatic chromosomal mosaicism. In: Liehr T (ed) Fluorescence in situ hybridization (FISH): application guide. Springer, Berlin, pp 301–311

Iourov IY, Vorsanova SG, Yurov YB (2010) Somatic genome variations in health and disease. Curr Genomics 11:387–396

Lemke J, Claussen J, Michel S, Chudoba I, Mühlig P, Westermann M, Sperling K, Rubtsov N, Grummt UW, Ullmann P, Kromeyer-Hauschild K, Liehr T, Claussen U (2002) The DNA-based structure of human chromosome 5 in interphase. Am J Hum Genet 71:1051–1059

Liehr T (ed) (2009) Fluorescence in situ hybridization (FISH): application guide. Springer, Berlin

Liehr T, Heller A, Starke H, Rubtsov N, Trifonov V, Mrasek K, Weise A, Kuechler A, Claussen U (2002) Microdissection-based high resolution multicolor banding for all 24 human chromosomes. Int J Mol Med 9:335–339

Liehr T, Weise A, Hinreiner S, Mkrtchyan H, Mrasek K, Kosyakova N (2010) Characterization of chromosomal rearrangements using multicolor-banding (MCB/m-band). Methods Mol Biol 659:231–238

Manvelyan M, Hunstig F, Bhatt S, Mrasek K, Pellestor F, Weise A, Simonyan I, Aroutiounian R, Liehr T (2008a) Chromosome distribution in human sperm: a 3D multicolor banding-study. Mol Cytogenet 1:25

Manvelyan M, Hunstig F, Mrasek K, Bhatt S, Pellestor F, Weise A, Liehr T (2008b) Position of chromosomes 18, 19, 21 and 22 in 3D-preserved interphase nuclei of human and gorilla and white hand gibbon. Mol Cytogenet 1:9

Manvelyan M, Kempf P, Weise A, Mrasek K, Heller A, Lier A, Höffken K, Fricke HJ, Sayer HG, Liehr T, Mkrtchyan H (2009) Preferred co-localization of chromosome 8 and 21 in myeloid bone marrow cells detected by three dimensional molecular cytogenetics. Int J Mol Med 24:335–341

Vorsanova SG, Yurov YB, Iourov IY (2010a) Human interphase chromosomes: a review of available molecular cytogenetic technologies. Mol Cytogenet 3:1

Vorsanova SG, Yurov YB, Soloviev IV, Iourov IY (2010b) Molecular cytogenetic diagnosis and somatic genome variations. Curr Genomics 11:440–446

Weise A, Starke H, Heller A, Uwe C, Liehr T (2002) Evidence for interphase DNA decondensation transverse to the chromosome axis: a multicolor banding analysis. Int J Mol Med 9:359–361

Yurov YB, Iourov IY, Vorsanova SG, Liehr T, Kolotii AD, Kutsev SI, Pellestor F, Beresheva AK, Demidova IA, Kravets VS, Monakhov VV, Soloviev IV (2007) Aneuploidy and confined chromosomal mosaicism in the developing human brain. PLoS One 2:e558

Yurov YB, Iourov IY, Vorsanova SG, Demidova IA, Kravetz VS, Beresheva AK, Kolotii AD, Monakchov VV, Uranova NA, Vostrikov VM, Soloviev IV, Liehr T (2008) The schizophrenia brain exhibits low-level aneuploidy involving chromosome 1. Schizophr Res 98:139–147

Yurov YB, Vorsanova SG, Iourov IY (2010) Ontogenetic variation of the human genome. Curr Genomics 11:420–425

Chapter 10
Chromosome Architecture Studied by High-Resolution FISH Banding in Three Dimensionally Preserved Human Interphase Nuclei

Thomas Liehr, Hasmik Mkrtchyan, Marina Manvelyan, Tatyana Karamysheva, Elisabeth Klein, and Samarth Bhatt

Abstract The impact of chromosome architecture on the formation of chromosome aberrations is a recent finding of interphase-directed molecular cytogenetic studies. Until recent years, biomedical research of interphase chromosomes in their integrity was hindered by technical limitations. The introduction of three-dimensional suspension-based fluorescence in situ hybridization (S-FISH) in combination with microdissection-based engineered DNA probes and fluorescence multicolor banding (MCB) allowed studying interphase chromosome organization, numbers, and rearrangements in different kinds of cells. Such studies have already provided comprehensive information on the interphase architecture of normal human sperm, as well as first insights into the influence of chromosomal rearrangements on the 3D

T. Liehr (✉) • E. Klein • S. Bhatt
Jena University Hospital, Institute of Human Genetics, Friedrich Schiller University, Kollegiengasse 10, Jena 07743, Germany
e-mail: i8lith@mti.uni-jena.de

H. Mkrtchyan
Jena University Hospital, Institute of Human Genetics, Friedrich Schiller University, Kollegiengasse 10, Jena 07743, Germany

Centre of Medical Genetics and Primary Health Care, Department of Medical Genetics, 2 Koryun Street, 0025 Yerevan, Armenia

M. Manvelyan
Jena University Hospital, Institute of Human Genetics, Friedrich Schiller University, Kollegiengasse 10, Jena 07743, Germany

Research Center of Maternal and Child Health Protection, Mashtots Avenue 22, 0002 Yerevan, Armenia

T. Karamysheva
Jena University Hospital, Institute of Human Genetics, Friedrich Schiller University, Kollegiengasse 10, Jena 07743, Germany

Laboratory of Morphology and Function of Cell Structure, Institute of Cytology and Genetics, Russian Academy of Sciences, Siberian Branch, Lavrentiev Avenue 10, Novosibirsk 630090, Russian Federation

Y.B. Yurov et al. (eds.), *Human Interphase Chromosomes: Biomedical Aspects*,
DOI 10.1007/978-1-4614-6558-4_10,

structure of sperm nuclei. Also, the influence of additional chromosomal fragments present in a nucleus was successfully visualized by S-FISH. Finally, S-FISH supported the idea that disease-specific chromosomal translocations could be the result of tissue-specific genomic organization. Overall, S-FISH combined with MCB, but also with other DNA probes, is a tool with high potential to resolve the influence of chromosomal imbalances or rearrangements on the interphase architecture, the latter being possibly a part of epigenetic cell regulation.

Introduction

In the interphase nucleus, chromosomes are located in specific regions, which are called "chromosome territories" (Cremer and Cremer 2001; Williams and Fisher 2003; Branco and Pombo 2006). Our own multicolor banding (MCB)-based studies revealed that the chromosome shape itself is not lost in the interphase nucleus and that one can even identify "interphase chromosomes" instead of only the chromosome territory, even irrespective of the cell-cycle phase (Weise et al. 2002; Lemke et al. 2002).

Both chromosome size and gene density have been discussed as having an important impact on the nuclear position of chromosomes. Small chromosomes preferentially locate close to the center of the nucleus, whereas large chromosomes can be found in the nuclear periphery (Sun et al. 2000; Bolzer et al. 2005). On the other hand, Croft et al. (1999) demonstrated a gene density-correlated radial arrangement of chromosomes in nuclei. Gene-dense and early-replicating chromatin can be found mainly in the central part of the nucleus, whereas gene-poor and later-replicating chromatin is located in the nuclear periphery (Croft et al. 1999). Interestingly, this nuclear topological arrangement is conserved during primate evolution (Manvelyan et al. 2008a).

Here we summarize the published applications of suspension-based fluorescence in situ hybridization (S-FISH) combined with FISH banding (Liehr et al. 2002, 2006), in particular, the most used approach, array-proven MCB (Weise et al. 2008).

S-FISH: The Method

Performing of a FISH experiment on human meta- and interphase cells after the air-drying method is a well-established approach; it is routinely done as a one- to multicolor FISH test (Liehr et al. 2004a). However, the air-drying procedure of chromosome preparation, leading to well-spread metaphases under appropriate conditions, leads at the same time to flattening of the originally spherical interphase nuclei. Thus, interphase architecture is hard to study reliably on such a preparation (Hunstig et al. 2009), even though some basic insights can also be gained using such material for FISH banding (Weise et al. 2002; Lemke et al. 2002).

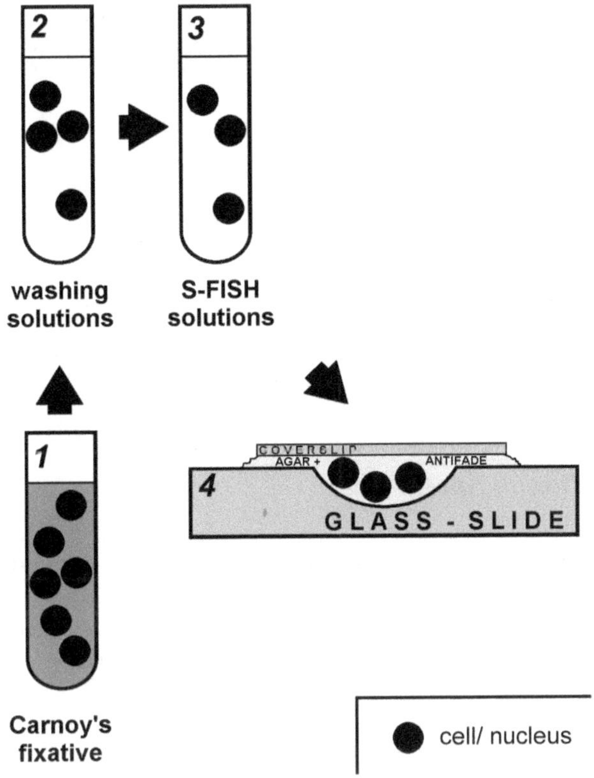

Fig. 10.1 Schematic drawing of the suspension-based fluorescence in situ hybridization (S-FISH) procedure. Overall, S-FISH avoids flattening and artificial swelling of the interphase nuclei, and the whole experiment is performed in suspension. A certain loss of cells during the washing steps is normal, shown here by the reduction of cells/nuclei from step 1 to 4. In principle, Carnoy's fixative is replaced subsequently by solutions necessary for a FISH, and washing steps are included as well. Finally, the cells/nuclei are immobilized and counterstained in agarose (AGAR) on a glass slide under a coverslip. Details of the protocol are described by Hunstig et al. (2009)

However, there is an easy way to do studies in three-dimensionally (3D) preserved interphase nuclei obtained from routinely prepared cytogenetic preparations stored in Carnoy's fixative. One only needs to do the whole FISH procedure in cell suspension, and as a final step the nuclei are placed on a polished concave slide before evaluation, immobilized in agarose. This approach for 3D-FISH analyses on totally spherical interphase nuclei, called suspension-based fluorescence in situ hybridization (S-FISH), was published first in 2002 (Steinhaeuser et al. 2002) and further developed and slightly modified later (Manvelyan et al. 2008a; Hunstig et al. 2009). Its principle is shown in Fig. 10.1.

S-FISH: Which DNA Probes May Be Applied?

For S-FISH, all available chromosomes or chromosome region-specific DNA are suitable. However, for application in S-FISH, at least double the amount of the probe is necessary compared to "normal" FISH experiments (Hunstig et al. 2009). To resolve the chromosome structure as a whole, single chromosome-directed FISH banding based on partial chromosome painting probes, as in MCB, is best suited

(Weise et al. 2008). Centromeric or locus-specific probes can be used as well for special questions (Manvelyan et al. 2009; Hunstig et al. 2009).

Applications of S-FISH

In addition to some studies done in comparative interphase cytogenetics of humans, the white-handed gibbon, and gorilla (Manvelyan et al. 2008a), S-FISH combined with MCB is mainly applied in the field of biomedical basic research of the human interphase nucleus. Here still many questions are open and unanswered, mainly because of the lack of suitable methods before introduction of S-FISH.

Human Sperm

For the first time the distribution of all human chromosomes in sperm was resolved comprehensively by S-FISH/MCB studies. Strikingly, for the majority of the 24 human chromosomes the distribution of the territories was alike as in lymphocytes; only the acrocentric chromosomes showed another location, because in sperm no nucleolus is formed (Manvelyan et al. 2008b). Thus, this nonrandom positioning must have biological significance. In other words, each chromosome needs to have a special position in the nucleus so that the cell can work properly. Sperm are translationally inactive cells; however, they need to have chromosomes at the right places as soon as a sperm enters an oocyte and needs to become active again.

The study by Manvelyan et al. (2008b) showed a direct correlation of chromosome positions and their sizes, apart from chromosomes 1, 2, 6, 14, 18, 20, 21, and Y; that is, large chromosomes were in the periphery, small ones in the center. Exactly those eight chromosomes not fitting in the previous correlation, perfectly aligned with gene density theory, that is, gene-dense chromosomes, were in the nuclear center and gene-poor ones in the periphery.

There is also already one study in sperm of males with a chromosomal aberration (Bhatt et al. 2009). Three males with paracentric inversion were studied, and no gross changes in the interphase positioning of the affected chromosomes were found. Certainly more studies on the influence of inborn rearrangements on the nuclear architecture of sperm, but also in other tissues, are necessary.

Different Tissues with Additional Chromosomal Fragments

Additional chromosomal material present in the cell is suspected to alter or at least influence the chromosomal architecture. Besides complete trisomies as inborn or acquired aberrations, there is the possibility of partial trisomies induced either by derivative chromosomes or by the presence of a small supernumerary marker chromosome (sSMC). The latter condition may be seen in 0.043 % of newborn infants, 0.077 % of prenatal cases, 0.433 % of mentally retarded patients, and 0.171 % of subfertile people (Liehr and Weise 2007). sSMC are defined as structurally abnormal chromosomes that cannot be identified or characterized unambiguously by conventional banding cytogenetics alone, and are generally equal in size or smaller than a chromosome 20 of the same metaphase spread. sSMC are mostly detected unexpectedly in routine cytogenetics (Liehr et al. 2004b). Also, they are not easy to correlate with a specific clinical outcome, because in addition to induction of genomic imbalance, mosaicism, etc., most often epigenetic factors can influence the phenotype of an sSMC carrier. Uniparental disomy, heterochromatization, and even their influence on the interphase architecture may play a role here.

In a recent study (Klein et al. 2012), S-FISH revealed that an extra piece of DNA such as an sSMC leads to gross rearrangements within the interphase nucleus, mainly concerning the sSMC normal sister chromosomes. Primarily, the position of the sSMC is influenced by or influences the position of the homologous chromosomes. sSMC and one sister chromosome tend to colocalize; this seems to be driven mainly by the amount of euchromatin present in the sSMC. Also, the sSMC seems to take over the position of one normal sister chromosome. Thus, the remainder sister chromosome is displaced toward another location within the nucleus. These observations were made in B- and T lymphocytes and skin fibroblasts.

Different Female Tissues and the Position of the X Chromosome

S-FISH/MCB studies in buccal mucosa, B- and T lymphocytes, and skin fibroblasts for the positioning of normal and derivative X chromosomes in female cells also may lead to interesting yet impossible insights into the nuclear architecture. Preliminary unpublished results (Fig. 10.2) first confirmed that active and inactive X chromosomes are located in the cell periphery and that the inactive X chromosome colocalizes to large parts, even though not perfectly, of the Barr body. Interestingly, a dicentric X chromosome, leading to an almost complete trisomy X, altered the positioning of the two X chromosomes to each other, inducing a larger distance between both normal and derivative X chromosomes compared to the normal cells. Thus, new insights may also be obtained by studying a well-known phenomenon such as X-inactivation by the S-FISH approach.

Leukemia and the Positions of Chromosomes 8 and 21

Nonrandom positioning of chromosomes in interphase nuclei is known to be of importance for genomic stability and formation of chromosome aberrations. Thus, tissue specificity of chromosomal translocations could be the result of tissue-specific

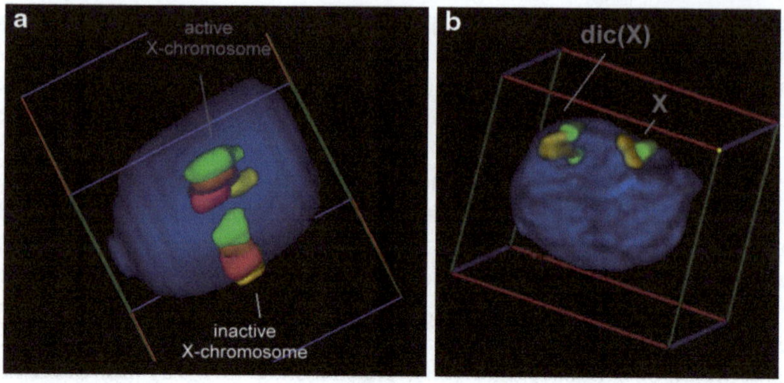

Fig. 10.2 S-FISH results after application of X chromosome-specific DNA probe sets. (**a**) Active and inactive X chromosomes in a lymphocyte nucleus of a normal female labeled with an MCB-X probeset. (**b**) A normal (X) and derivative X chromosome (dic(X)) labeled with partial chromosome paints for Xp (*green*) and Xq (*yellow*) visualized in the fibroblast cell line GM15859 (*Corriell*). The female carrier had a constitutional karyotype 46,X,dic(X)(pter->q28::q28->pter)

genome organization (Meaburn et al. 2007; Brianna Caddle et al. 2007), and a positive correlation between spatial proximity of chromosomes/genes in interphase nuclei and translocation frequencies was shown (Bickmore and Teague 2002; Roix et al. 2003; Branco and Pombo 2006; Meaburn et al. 2007; Brianna Caddle et al. 2007; Grasser et al. 2008).

Manvelyan et al. (2009) provided evidence that there might be an effect of specific chromosome positioning in myeloid bone marrow cells, that is, a colocalization of chromosomes 8 and 21 could promote a translocation providing selective advantage of t(8;21) cells in AML-M2. Yet unpublished additional S-FISH studies confirmed that this is specifically true for AML patients having a trisomy 8. Overall, studies to enlighten the nuclear position of tumor-related oncogenes, which are known to be activated by specific translocations, are promising targets of future S-FISH-studies, as supported by recent comparable findings in thyroid cancer (Gandhi et al. 2009).

S-FISH, Conclusions, and Perspectives

Overall, the combination of S-FISH and MCB for a three-dimensional analysis of chromosome position in the interphase nucleus, which can be accompanied by the use of locus-specific probes, is a powerful tool. The topological organization in interphase nucleus is nonrandom, and it becomes more and more obvious that there is a physiological reason behind that.

The completed S-FISH studies in humans that have been already summarized show the potential of this approach for (1) genome-wide analysis of interphase architecture in tissues not yet studied (such as done for sperm; Manvelyan et al.

2008b), (2) studies on architectural changes in nuclei with additional chromosomes or chromosomal material (such as done for sSMC; Klein et al. 2012, or the X chromosome), and (3) analysis for the susceptibility of specific parts of the human genome for rearrangements caused by colocalization (as done for the t(8;21) in AML; Manvelyan et al. 2009). It is certain that additional biomedical research aspects of interphase chromosomes may also be discovered using the S-FISH/MCB approach.

Acknowledgments Work supported in part by the DFG (436 ARM 17/11/06, LI 820/15-1, LI 820/21-1, LI 820/24-1, LI 820/33-1), DAAD (A/07/04616), Stefan-Morsch-Stiftung, Monika Kutzner Stiftung, Else Kröner-Fresenius-Stiftung (2011_A42).

References

Bhatt S, Moradkhani K, Mrasek K, Puechberty J, Manvelyan M, Hunstig F, Lefort G, Weise A, Lespinasse J, Sarda P, Liehr T, Hamamah S, Pellestor F (2009) Breakpoint mapping and complete analysis of meiotic segregation patterns in three men heterozygous for paracentric inversions. Eur J Hum Genet 17:44–50

Bickmore WA, Teague P (2002) Influences of chromosome size, gene density and nuclear position on the frequency of constitutional translocations in the human population. Chromosome Res 10:707–715

Bolzer A, Kreth G, Solovei I, Koehler D, Saracoglu K, Fauth C, Muller S, Eils R, Cremer C, Speicher MR, Cremer T (2005) Three-dimensional maps of all chromosomes in human male fibroblast nuclei and prometaphase rosettes. PLoS Biol 3:e157

Branco MR, Pombo A (2006) Intermingling of chromosome territories in interphase suggests role in translocations and transcription-dependent associations. PLoS Biol 4:e138

Brianna Caddle L, Grant JL, Szatkiewicz J, van Hase J, Shirley BJ, Bewersdorf J, Cremer C, Arneodo A, Khalil A, Mills KD (2007) Chromosome neighborhood composition determines translocation outcomes after exposure to high-dose radiation in primary cells. Chromosome Res 15:1061–1073

Cremer T, Cremer C (2001) Chromosome territories, nuclear architecture and gene regulation in mammalian cells. Nat Rev Genet 2:292–301

Croft JA, Bridger JM, Boyle S, Perry P, Teague P, Bickmore WA (1999) Differences in the localization and morphology of chromosomes in the human nucleus. J Cell Biol 45:1119–1131

Gandhi MS, Stringer JR, Nikiforova M, Medvedovic M, Nikiforov YE (2009) Gene position within chromosome territories correlates with their involvement in distinct rearrangement types in thyroid cancer cells. Genes Chromosomes Cancer 48:222–228

Grasser F, Neusser M, Fiegler H, Thormeyer T, Cremer M, Carter NP, Cremer T, Miller S (2008) Replication-timing correlated spatial chromatin arrangements in cancer and in primate interphase nuclei. J Cell Sci 121:1876–1886

Hunstig F, Manvelyan M, Bhatt S, Steinhaeuser U, Liehr T (2009) Three-dimensional interphase analysis enabled by suspension FISH. In: Liehr T (ed) Fluorescence in situ hybridization (FISH)—application guide, 1st edn. Springer, Berlin

Klein E, Manvelyan M, Simonyan I, Hamid AB, Santos Guilherme R, Liehr T, Karamysheva T (2012) Centromeric association of small supernumerary marker chromosomes with their sister-chromosomes detected by three dimensional molecular cytogenetics. Mol Cytogenet 5:15

Lemke J, Claussen J, Michel S, Chudoba I, Mühlig P, Westermann M, Sperling K, Rubtsov N, Grummt UW, Ullmann P, Kromeyer-Hauschild K, Liehr T, Claussen U (2002) The DNA-based structure of human chromosome 5 in interphase. Am J Hum Genet 71:1051–1059

Liehr T, Weise A (2007) Frequency of small supernumerary marker chromosomes in prenatal, newborn, developmentally retarded and infertility diagnostics. Int J Mol Med 19:719–731
Liehr T, Heller A, Starke H, Claussen U (2002) FISH banding methods: applications in research and diagnostics. Expert Rev Mol Diagn 2:217–225
Liehr T, Starke H, Weise A, Lehrer H, Claussen U (2004a) Multicolor FISH probe sets and their applications. Histol Histopathol 19:229–237
Liehr T, Claussen U, Starke H (2004b) Small supernumerary marker chromosomes (sSMC) in humans. Cytogenet Genome Res 107:55–67
Liehr T, Starke H, Heller A, Kosyakova N, Mrasek K, Gross M, Karst C, Steinhaeuser U, Hunstig F, Fickelscher I, Kuechler A, Trifonov V, Romanenko SA, Weise A (2006) Multicolor fluorescence in situ hybridization (FISH) applied to FISH-banding. Cytogenet Genome Res 114:240–244
Manvelyan M, Hunstig F, Mrasek K, Bhatt S, Pellestor F, Weise A, Liehr T (2008a) Position of chromosomes 18, 19, 21 and 22 in 3D-preserved interphase nuclei of human and gorilla and white hand gibbon. Mol Cytogenet 1:9
Manvelyan M, Hunstig F, Bhatt S, Mrasek K, Pellestor F, Weise A, Simonyan I, Aroutiounian R, Liehr T (2008b) Chromosome distribution in human sperm—a 3D multicolor banding-study. Mol Cytogenet 1:25
Manvelyan M, Kempf P, Weise A, Mrasek K, Heller A, Lier A, Höffken K, Fricke HJ, Sayer HG, Liehr T, Mkrtchyan H (2009) Preferred co-localization of chromosome 8 and 21 in myeloid bone marrow cells detected by three dimensional molecular cytogenetics. Int J Mol Med 24:335–341
Meaburn KJ, Misteli T, Soutoglou E (2007) Spatial genome organization in the formation of chromosomal translocations. Semin Cancer Biol 17:80–90
Roix JJ, McQueen PG, Munson PJ, Parada LA, Misteli T (2003) Spatial proximity of translocation-prone gene loci in human lymphomas. Nat Genet 34:287–291
Steinhaeuser U, Starke H, Nietzel A, Lindenau J, Ullmann P, Claussen U, Liehr T (2002) Suspension (S)-FISH, a new technique for interphase nuclei. J Histochem Cytochem 50:1697–1698
Sun HB, Shen J, Yokota H (2000) Size-dependent positioning of human chromosomes in interphase nuclei. Biophys J 79:184–190
Weise A, Starke H, Heller A, Claussen U, Liehr T (2002) Evidence for interphase DNA decondensation transverse to the chromosome axis: a multicolor banding analysis. Int J Mol Med 9:359–361
Weise A, Mrasek K, Fickelscher I, Claussen U, Cheung SW, Cai WW, Liehr T, Kosyakova N (2008) Molecular definition of high-resolution multicolor banding probes: first within the human DNA sequence anchored FISH banding probe set. J Histochem Cytochem 56:487–493
Williams RE, Fisher AG (2003) Chromosomes, positions please! Nat Cell Biol 5:388–390

Chapter 11
Technological Solutions in Human Interphase Cytogenetics

Svetlana G. Vorsanova, Yuri B. Yurov, and Ivan Y. Iourov

Abstract Numerous interphase molecular cytogenetic approaches are useful for the analysis of chromosomes in normal and abnormal human cells. Interphase fluorescence in situ hybridization techniques offer unique possibilities to visualize individual chromosomes or chromosomal regions in single nondividing cells isolated from any given tissue. Despite technological difficulties encountered during studying human interphase chromosomes in health and disease, molecular cytogenetics or cytogenomics ("chromosomics") does provide solutions for high-resolution single-cell analysis of genome organization, structure, and behavior at all stages of the cell cycle. However, usually relatively little attention is paid to interphase molecular cytogenetics in current biomedical literature. Looking through the voluminous amount of original research papers and reviews dedicated to human interphase chromosomes, one can conclude that the technological aspects of studying human interphase chromosomes applied to basic and clinical research are rarely addressed. In an attempt to fill this gap, the present chapter provides a description of technological solutions in human interphase cytogenetics.

S.G. Vorsanova (✉) • Y.B. Yurov
Mental Health Research Center, Russian Academy of Medical Sciences, Moscow, Russia

Institute of Pediatrics and Children Surgery, Ministry of Health of Russian Federation, Moscow 127412, Russia

Moscow City University of Psychology and Education, Moscow, Russia
e-mail: svorsanova@mail.ru; y_yurov@yahoo.com

I.Y. Iourov
Mental Health Research Center, Russian Academy of Medical Sciences, Moscow, Russia

Institute of Pediatrics and Children Surgery, Ministry of Health of Russian Federation, Moscow 127412, Russia
e-mail: ivan.iourov@gmail.com

Y.B. Yurov et al. (eds.), *Human Interphase Chromosomes: Biomedical Aspects*,
DOI 10.1007/978-1-4614-6558-4_11, © Springer Science+Business Media, LLC 2013

Introduction

It is generally accepted that almost all fluorescence in situ hybridization (FISH) protocols are applicable for developing an interphase FISH (I-FISH) method. To learn more about numerous approaches and applications of useful FISH-based techniques and detailed protocols, readers can refer to recent FISH application guides edited by the authors with the contribution of leading international experts in the field of molecular cytogenetics and cytogenomics, who also contributed to this book (*Fluorescence In Situ Hybridization (FISH)—Application Guide*, edited by Thomas Liehr, Springer, 2009, and *Fluorescence In Situ Hybridization (FISH): Protocols and Applications*, edited by Joanna M. Bridger and Emanuela V. Volpi, Humana Press, 2010).

Recently we have attempted to give an overview of currently applied molecular cytogenetic techniques with a special emphasis on their technological abilities for studying human interphase chromosomes (Vorsanova et al. 2010a, freely available at Molecular Cytogenetics (BioMed Central), an open access journal dedicated to different aspects of chromosome and genome biology, http://www.molecularcytogenetics.org/content/3/1/1). Here we present an updated review dedicated to technological achievements in human interphase cytogenetics.

According to Gersen and Keagle (2005), it is estimated that more than one million cytogenetic and molecular cytogenetic analyses are performed each year. Taken together, these analyses represent the standard of care in medical genetics and routine clinical workups for numerous patients suffering from congenital malformations, mental diseases, cancers, or reproductive problems (Carter 2007; Liehr 2009; Vorsanova et al. 2010b). The significance of molecular cytogenetic techniques for molecular diagnosis has been repeatedly shown, and these techniques are recognized as a valuable addition or even alternative to conventional cytogenetics (Liehr and Claussen 2002; Iourov et al. 2008a; Bejjani and Shaffer 2008). In addition, molecular cytogenetic technologies are widely used in basic biomedical research (Liehr et al. 2004). For instance, the thousands of articles mentioning at least one molecular cytogenetic technique are indexed in browsable scientific databases (for more details, see Iourov et al. 2008a; Chap. 12, and the web page about multicolor FISH at http://www.med.uni-jena.de/fish/mFISH/mFISHlit.htm website, managed by Dr. Thomas Liehr, Jena, Germany). Thus, it is certain that the role of molecular cytogenetics in current biomedicine is appreciable.

Two essential advantages of molecular cytogenetics can be noted: (1) the ability to provide either a high-resolution on-chip scan of the whole genome or to visualize single specific genomic loci (Bejjani and Shaffer 2008), and (2) the capability to analyze DNA (RNA)-based genome organization, structure, and behavior in single cells (Levsky and Singer 2003; Iourov et al. 2006a). The first advantage is appreciable when analyzing mixed DNA isolated from a large amount of cells and is rarely appreciated in single-cell genomic studies (Iourov et al. 2012; Vanneste et al. 2012). The second advantage of molecular cytogenetic techniques is consistently emphasized but is usually applied to studying metaphase chromosomes of mitotic cells.

However, eukaryotic cells are more likely to be in interphase. Therefore, surveys of genome organization, structure, and behavior do not evaluate an essential part of cellular life. In molecular diagnosis, interphase analysis is also not commonly applied. One might suggest a lack of reproducibility and the low resolution of interphase cytogenetic techniques. However, a brief look through molecular cytogenetic studies of somatic genomic variations and genome behavior in interphase nuclei (Walter et al. 2006; Goetze et al. 2007; Iourov et al. 2008b, 2012; Rouquette et al. 2010) and developments in interphase cytogenetics (Iourov et al. 2006c; Liehr 2009; Vorsanova et al. 2010a, b) demonstrates that this idea is unsupported. Furthermore, numerous laboratories elaborating such techniques are able to solve important practical and research tasks without notable difficulties. Evidently, interphase molecular cytogenetics requires additional attention, which is the intention of the present chapter.

Molecular Cytogenetic Techniques and Interphase Cytogenetics

There are currently two essential platforms available for developments in molecular cytogenetics: FISH, including comparative genomic hybridization (CGH), and peptide nucleic acid (PNA) probing for analysis of chromosomal DNA (Liehr and Claussen 2002; Iourov et al. 2008a; Liehr 2009; Vorsanova et al. 2010a). The resolution of such techniques is usually established against cytogenetic banding analysis (the gold standard of resolution for genetic analyses). Single-cell molecular cytogenetics analyzes either metaphase plates or interphase nuclei. Study of metaphase plates is traditionally made by means of several detection technologies [spectral karyotyping (SKY) or multicolor FISH (MFISH)] (Schrock et al. 1996; Speicher et al. 1996) or specific DNA probe sets (chromosome-enumeration/centromeric, site-specific, whole-painting, microdissected) (Yurov et al. 1996; Soloviev et al. 1998a, b; Liehr et al. 2002; Nietzel et al. 2001). If modified, these techniques can be applied to interphase chromosomal analysis, but this "translation" (transfer of technology) requires major efforts (Vorsanova et al. 2010a; Iourov et al. 2010). Table 11.1 provides an overview of the molecular cytogenetic techniques used for metaphase and interphase analysis of single cells. As one can see, molecular cytogenetics allows us to perform high-resolution analysis of chromosomal structure and behavior at all stages of the cell cycle. Nonetheless, molecular cytogenetic methods are preferentially used for detecting metaphase chromosome imbalances and rearrangements or for whole-genome scans by CGH (Liehr et al. 2004; Vorsanova et al. 2010b).

Visualization is the key stage of studying interphase chromosomes. Without direct (microscopic) visualization of DNA-base chromosomal structures, related research is certainly incomplete. Thus, FISH-based techniques offer the unique possibility to depict either whole chromosomes or specific genomic loci in single cells. In other words, if one wishes to obtain valid data on human interphase chromosomes, one will undertake an I-FISH study. Further, we attempt to review the

Table 11.1 Molecular cytogenetic techniques

		Resolution	MA[a]	IA[b]	SCA[c]	VC[d]
Conventional cytogenetics (banding)		>5–7 Mb	+	–	+	+
FISH/MFISH/SKY	With centromeric probes	>0.3–1 Mb	+/–	+	+	+/–
	With site-specific probes	~0.1–2 Mb	+/–	+/–	+/–	+/–
	With whole-painting probes	>5–10 Mb	+	–	+	+
MCB	Metaphase MCB	~2–5 Mb	+	–	+	+
	ICS-MCB	~2–5 Mb	–	+	+	+
	Fiber FISH	>3 kb	na	na	+	+
Single-cell CGH	Standard CGH	>5 Mb	na	na	+	–
	Array CGH	>0.03 Mb	na	na	+	–

Source: Adapted from Vorsanova et al. (2010a)
FISH fluorescence in situ hybridization, *MFISH* multicolor FISH, *SKY* spectral karyotyping, *MCB* multicolor banding, *CGH* comparative genomic hybridization, *na* not applicable
[a]Analysis of metaphase chromosomes (*MA* metaphase analysis)
[b]Analysis of interphase chromosomes (*IA* interphase analysis)
[c]Possibility to perform single cell-analysis (SCA)
[d]Possibility to visualize chromosomes or chromosomal loci (*VC* visualizing chromosomal loci)

technique in context of applications to single-cell chromosomal analysis, which is the basis of interphase molecular cytogenetics.

I-FISH

FISH represents a general-purpose technique for studies of both the whole genome and specific genomic loci. The resolution of molecular cytogenetic is essentially determined according to the DNA sequence size of probes hybridizing in situ. DNA probes are designated as centromeric and telomeric (hybridizing to repetitive-sequence DNA), site-specific (hybridizing to euchromatic DNA, i.e., DNA sequences of a gene), or whole chromosome painting (wcp; hybridizing to DNA of whole chromosomes) (Table 11.2).

FISH with chromosome-specific DNA probes: FISH painting of repetitive genomic sequences is performed with centromeric (chromosome-enumeration or chromosome-specific) or telomeric DNA probes. Analysis of telomeres is an important area of biomedical research (Aubert et al. 2012). In such approaches, PNA/DNA probes possessing TTAGGG repetitive motifs are used. Representing the technological basis of telomere biology (cancer and aging research), telomere FISH and related techniques are poorly applicable for diagnosis. On the other hand, I-FISH with telomeric probes is applicable for analysis of nuclear organization (Klewes et al. 2011).

I-FISH using centromeric DNA probes has become an integral part of molecular diagnosis in medical genetics, oncology, and reproductive medicine (Cremer et al. 1986; Vorsanova et al. 1986, 1991, 2005b; Baumgartner et al. 2006; Yurov et al. 2007; Iourov et al. 2008b). Additionally, it is repeatedly demonstrated that I-FISH

Table 11.2 Interphase FISH (I-FISH) overview

Technique	Description	Advantages	Disadvantages
I-FISH with centromeric probes	Painting of pericentromeric (heterochromatic) regions	High hybridization efficiency, chromosome specificity (apart from chromosomes 5/19, 13/21, 14/22)	Chromosomal associations (associations of signals causing misinterpretation of false-positive monosomy), impossibility of analyzing chromosomes 5, 13, 14, 19, 21, 22; heteromorphisms
I-FISH with site-specific probes	Painting of specific euchromatic genomic loci	Specific genomic loci (»1 Mb) are visualized	Low hybridization efficiency, numerous artifacts
I-FISH with wcp	I-FISH painting chromosome territories	Visualization of chromosome territories in interphase nuclei	Chromosome territories are ambiguous, additional information about behavior of chromosomal regions is occasional
mFISH	Multicolor I-FISH with several differentially labeled probes	Analysis of several targeted genomic loci	Difficulty to distinguish between artifacts and chromosomal abnormalities
mFISH/QFISH	mFISH+QFISH (QFISH: digitalization of FISH signals)	Possibility to distinguish between FISH artifacts and numerical chromosomal imbalances	Same as mFISH
MFISH	Visualization of the complete set of chromosomes in an interphase nucleus	All chromosome territories are simultaneously seen	Exceedingly sophisticated analysis; data poorly interpretable
ICS-MCB	Chromosome-specific MCB generated on interphase nuclei	Visualization of whole banded interphase chromosomes in their integrity	Single pair of homologous chromosomes is visualized per assay; relative complexity of the analysis

Source: Adapted from Vorsanova et al. (2010a)

with these probes is highly applicable in chromosome biology studies encompassing genome research (chromosomal and nuclear), evolution, behavior, and variation in health and disease. These DNA probes feature nearly 100 % hybridization efficiency and chromosome specificity. As a result, analysis of an individual homologous chromosome pair in interphase becomes possible. Moreover, extreme interindividual variations of pericentromeric heterochromatic DNA has led to developing quantitative FISH (QFISH) solving numerous problems encountered during metaphase and interphase analysis of chromosomes (Iourov et al. 2005;

Vorsanova et al. 2005a). It is noteworthy that the resolution of related assays is poorly determined by DNA sequence size of loci assessed (Table 11.1), inasmuch as centromeric I-FISH is used for the analysis of phenomena involving large chromosomal regions or even whole chromosomes. For instance, I-FISH with chromosome-enumeration probes allows the detection of numerical chromosome imbalances (aneuploidy and polyploidy) in interphase. The latter is the most frequent application of the method (Fig. 11.1) and is required for pre-/postnatal diagnosis, cancer diagnosis/prognosis, and somatic genomic variation surveys. The nearly 100 % hybridization efficiency of centromeric DNA probes and chromosome-specific DNA sequences forming pericentromeric (heterochromatic) chromosomal regions (apart from shared alphoid DNA of chromosomes 5 and 19, 13 and 21, and 14 and 22) (Yurov et al. 1996; Lee et al. 1997; Vorsanova et al. 2002, 2005a) is the essential advantage that this technique possesses. However, heteromorphisms of pericentromeric DNAs can produce the lack of a signal leading, thereby, to impossibility of the I-FISH assay application. Fortunately, such extreme heteromorphisms (centromeric DNA variations) are rare in the general population (Verma and Luke 1992; Liehr et al. 1998; Vorsanova et al. 2002, 2005a).

I-FISH with site-specific probes: Site-specific DNA probes (YACs, BACs, PACs, cosmids) are used either to map chromosomal regions within which a breakpoint is located or to evaluate chromosomal imbalances by a targeted FISH assay (diagnosis of known microdeletion and microduplication syndromes) (Iourov et al. 2008a; Liehr 2009), aneuploidy and/or recurrent chromosome abnormalities during preimplantation genetic diagnosis (Fung et al. 2001; Stumm et al. 2006; Lu et al. 2009), prenatal diagnosis (Soloviev et al. 1995; Vorsanova et al. 2005b; Liehr 2009), oncocytogenetic analysis (Liehr and Claussen 2002; Mitelman et al. 2007), and copy number variation precision (Carter 2007). Probing small genomic loci (<1 Mb), site-specific probes are applied to studying gene-specific nuclear organization and its relevance to genome behavior (Goetze et al. 2007; Strickfaden et al. 2010). However, relatively moderate hybridization efficiency (<70 %) hinders using the approaches in numerous areas of biomedical research and diagnosis. Alternatively, a number of FISH procedures with these types of probes (i.e., hematological and tumor diagnosis) are found effective for molecular cytogenetic diagnosis and have cutoffs varying between 90 and 95 % (Liehr 2009). As a result, interphase molecular cytogenetic studies by I-FISH with site-specific probes are commonly applied in preimplantation, prenatal, and postnatal diagnosis as well as in cancer cytogenetics (Fig. 11.2). Although repeatedly noted to be of significant importance for detecting gene fusions resulting from interchromosomal translocations (cancer biomarkers) (Mitelman et al. 2007) and to be useful for preimplantation diagnosis (Stumm et al. 2006), such I-FISH modifications have considerable disadvantages. To be more precise, the hybridization efficiency of site-specific probes is usually between 40 % and 70 %. This irregularity of hybridization efficiency can produce false-positive or false-negative data. Moreover, one has to use probes hybridizing to well-characterized chromosomal/genomic DNA loci (i.e., oncogenes or genes/genomic loci within deletion or duplication regions). Few well-characterized approaches using these DNA probes may be of importance for detecting continuously reported

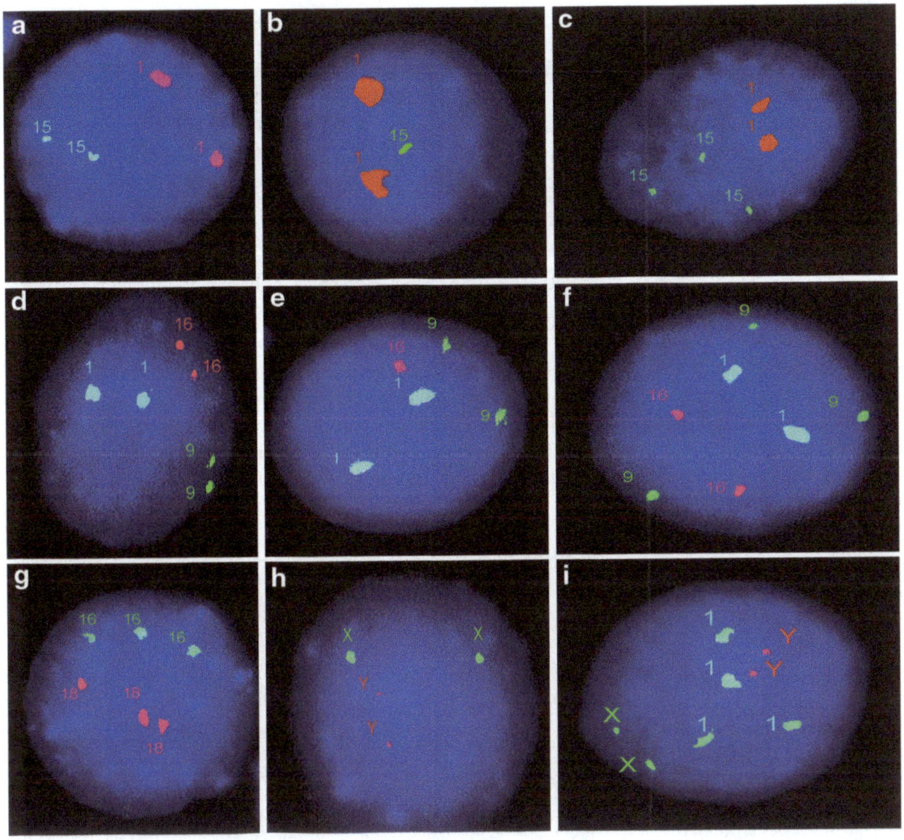

Fig. 11.1 Two- and three-color interphase fluorescence in situ hybridization (I-FISH) with centromeric DNA probes: (**a**) normal diploid nucleus with two signals for chromosome 1 and chromosome 15; (**b**) monosomic nucleus with two signals for chromosome 1 and one signal for chromosome 15; (**c**) trisomic nucleus with two signals for chromosome 1 and three signals for chromosome 15; (**d**) normal diploid nucleus with two signals for chromosome 1, chromosome 9, and chromosome 16; (**e**) monosomic nucleus with two signals for chromosome 1 and chromosome 9 and one signal for chromosome 16; (**f**) trisomic nucleus with two signals for chromosome 1 and chromosome 16 and three signals for chromosome 9; (**g**) triploid nucleus with three signals for chromosome 16 and chromosome 18; (**h**) tetraploid nucleus with two signals for chromosome X and chromosome Y; (**i**) tetraploid nucleus with two signals for chromosome X and chromosome Y and four signals for chromosome 1. (Copyright © Vorsanova et al. 2010a; licensee BioMed Central Ltd. This is an Open Access article distributed under the terms of the Creative Commons Attribution License (http://creativecommons.org/licenses/by/2.0)

chromosomal rearrangements in cancer cells (Virgili et al. 2008; Nicholson and Duesberg 2009; Sen and Hopwood 2010) and deletions/duplications in a clinical population (Halder et al. 2008; Weise et al. 2012). Nonetheless, application of site-specific probes is the best way to visualize interphase chromosomal DNAs less than 1 Mb. Simultaneous hybridization of centromeric and site-specific probes (mFISH) (Fig. 11.3) is applicable for diagnostics and survey of somatic genome variations.

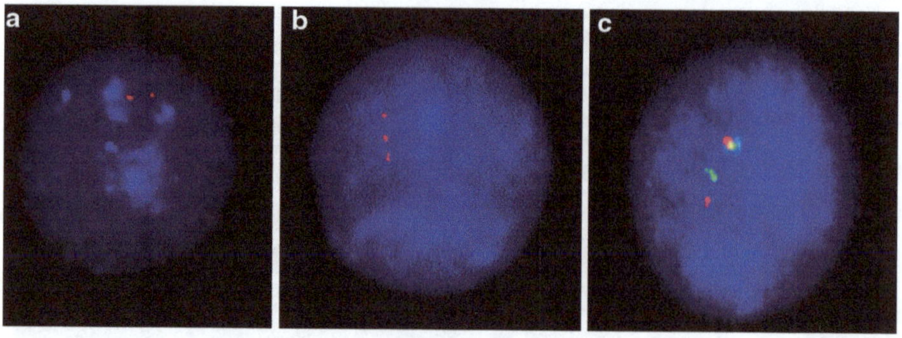

Fig. 11.2 I-FISH with site-specific DNA probes: (**a**) normal diploid nucleus with two signals for chromosome 21; (**b**) trisomic nucleus with three signals for chromosome 21; (**c**) interphase nucleus exhibiting colocalization of *ABL* and *BCR* genes, probably caused by t(9;22)/Philadelphia chromosome. (Copyright © Vorsanova et al. 2010a; licensee BioMed Central Ltd. This is an Open Access article distributed under the terms of the Creative Commons Attribution License (http://creativecommons.org/licenses/by/2.0)

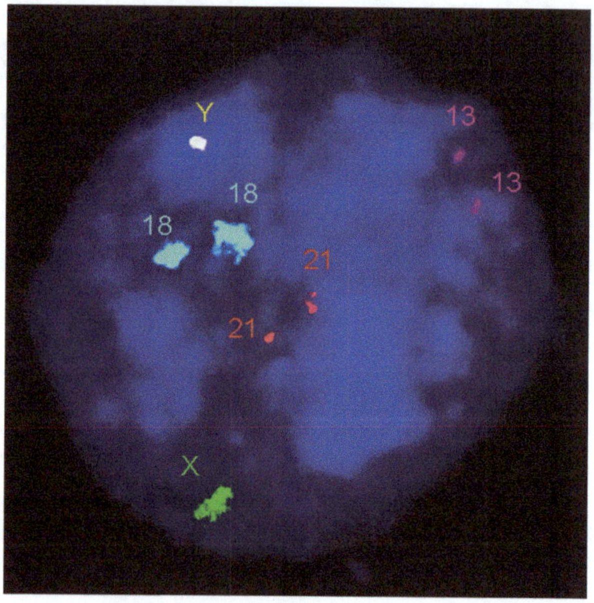

Fig. 11.3 Five-color I-FISH (mFISH) with DNA probes for chromosomes 18, X, and Y (centromeric probes) as well as 13 and 21 (site-specific probes). A presumably normal (diploid) male nucleus isolated from the adult human brain. (Copyright © Vorsanova et al. 2010a; licensee BioMed Central Ltd. This is an Open Access article distributed under the terms of the Creative Commons Attribution License (http://creativecommons.org/licenses/by/2.0)

FISH with wcp probe and MFISH/SKY: Wcp probe combinations allow performing 24-color MFISH or SKY for metaphase analysis of interchromosomal chromosome rearrangements in cancers and individuals with constitutional chromosomal abnormities (Liehr and Claussen 2002; Liehr et al. 2004). In interphase chromosomal analysis, MFISH and SKY are hardly applicable. Occasional studies applied MFISH-based approaches for visualizing all chromosomes in fibroblast interphase nuclei and prometaphase rosettes (Walter et al. 2006). Similar assays with 2–5 wcp probes are frequently encountered in molecular cytogenetic diagnosis of structural

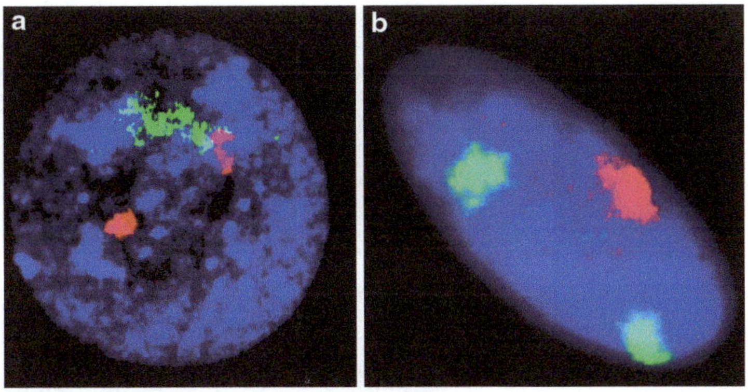

Fig. 11.4 I-FISH with two-whole chromosome painting (wcp) for chromosomes 7 and 21. (**a**) Ambiguous chromosome territories provide information neither about number of chromosomes nor about structure of chromosomes (chromosome 7, *green signal*; chromosome 21, *red signals*), whereas this individual presented with regular unbalanced t(7;21). Details of this case are described in Vorsanova et al. (2008). (**b**) Chromosome territories in an interphase nucleus of a cell isolated from the ataxia-telangiectasia brain (chromosome 7, *green signals*; chromosome 14, *red signal*). Note the impossibility to identify number of chromosomes 14. (Copyright © Vorsanova et al. 2010a; licensee BioMed Central Ltd. This is an Open Access article distributed under the terms of the Creative Commons Attribution License (http://creativecommons.org/licenses/by/2.0)

alterations to metaphase chromosomes, as well (Liehr et al. 2004), and in investigation of genome organization in interphase nuclei (Rouquette et al. 2010). Nevertheless, I-FISH with wcp probes is sophisticated and too poorly reproducible to become competitive with other interphase molecular cytogenetic techniques. It is therefore unsurprising that FISH chromosomal painting by wcp is generally recognized as completely useless for identification of interphase chromosome numbers and structure (Fig. 11.4). Alternatively, basic research of chromosome architecture in interphase is usually performed using I-FISH with wcp providing for visualization of chromosome territories and their positioning relative to nuclear structures. Additionally, I-FISH-wcp approaches were almost the only way to study genomic organization in interphase until more effective techniques have been elaborated (Walter et al. 2006; Rouquette et al. 2010). Finally, these techniques are all limited in their abilities to paint chromosome territories (volumes) only (Table 11.2).

Interphase chromosome-specific MCB: By microdissection of chromosomal loci for obtaining a set of probes that produce multicolor pseudo-G-banding, a high-resolution molecular cytogenetic technique for analysis of metaphase chromosomes termed MCB (multicolor banding) was proposed (Liehr et al. 2002). Further, this idea has been adapted to interphase chromosomal analysis and has provided for elaboration of interphase chromosome-specific MCB (ICS-MCB). To visualize a homologous pair of interphase chromosomes in their integrity, one has to generate MCB (Iourov et al. 2007, 2009a, b; Manvelyan et al. 2008; Iourov 2012). Figure 11.5 gives an example of aneuploidy detection in an interphase nucleus isolated from the Alzheimer's disease brain (Iourov et al. 2009a). ICS-MCB can be widely applied

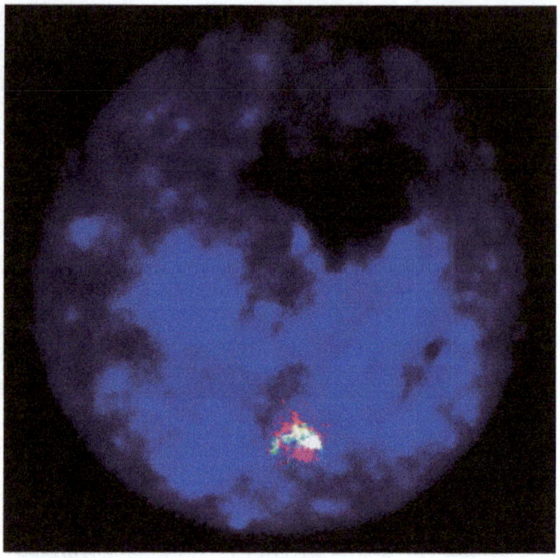

Fig. 11.5 Interphase chromosome-specific multicolor banding (ICS-MCB) with chromosome 21-specific probe. Monosomy (loss) of chromosome 21 in a nucleus isolated from the Alzheimer's disease brain. (Copyright © Vorsanova et al. 2010a; licensee BioMed Central Ltd. This is an Open Access article distributed under the terms of the Creative Commons Attribution License (http://creativecommons.org/licenses/by/2.0)

for basic research of somatic genomic variations, chromosome structural and functional organization in interphase, and supramolecular disease mechanisms. Apparently, the sole disadvantage of this technique is the impossibility to analyze more than one homologous chromosome pair at a time (Iourov et al. 2007; Iourov 2012). This state-of-the-art molecular cytogenetic technology is discussed in detail in Chap. 9 of this book.

Fiber FISH: Probably the highest molecular cytogenetic resolution is achieved by DNA fiber FISH. This technique provides a mapping resolution of 1–3 Mb (metaphase analysis). Applied to interphase nuclei, it achieves a resolution of 50 kb or even more. The high resolution is attributed to a higher degree of chromatin decondensation than other FISH techniques. Applied to naked DNA fibers, fiber FISH show that chromatin fully decondenses, leading to a resolution ranging from 1 to 400 kb. Furthermore, DNA fiber FISH provides a mapping tool supplementary to restriction mapping permitting accurate gap and overlap sizing (Raap et al. 1996; Weier 2001). The latter, however, is currently out of the scope of human genome research, inasmuch as genomic loci are supposed to be all mapped in a definitive manner by the Human Genome Project.

Heng et al. (1992) were able to release the chromatin fibers from cells arrested at G_1 and G_2 by different chemicals and alkaline lysis procedure. They have also demonstrated fluorescence-labeled probes to hybridize specifically to single-copy genomic DNA sequences of the free chromatin. FISH signals have been detected for sequences separated by 21 kb (the closest position) and 350 kb (the far position), with exact correspondence between the observed and expected distances. The resolution of this technique is likely to approach 10 kb, and the coverage should span millions of base pairs. According to these data, authors have concluded that free chromatin mapping can be generally used to study the structure and organization of mammalian interphase genomes.

To improve the DNA resolution of FISH, Wiegant and his collaborators have adapted a nuclear extraction technique, resulting in highly extended DNA loops arranged around the nuclear matrix in a halo-like structure (Wiegant et al. 1992). In situ hybridization signals depicting alphoid and cosmid DNAs appear as beads-on-a-string, which, according to preliminary experiments, results from the association of individual probe fragments. By multicolor hybridization the authors were able to determine relative map positions and to detect easily a 10-kb overlap between individual cosmid clones, each of which shows linear beaded signals, suggesting that the DNA was essentially linearized in these experiments. The resolution range was defined as 10–200 kb, and probably as little as a few kilobases, thus greatly extending the abilities of fiber FISH. Fiber FISH was also found useful for investigation of genomic organization and mapping, stalled transcription, and genomic rearrangements (including large deletions within gene sequences) (Weier 2001). Although this technique is based on obtaining DNA fibers from interphase nuclei, it cannot be directly attributed to I-FISH. Single-cell fiber FISH (especially when large cell populations are analyzed) is highly complicated.

Immuno-FISH: Immuno-FISH combines immunohistochemical detection of proteins with FISH to specific DNA (RNA) targets (Dundas et al. 2001; Yang et al. 2004). A simple protocol of immuno-FISH using cytospin centrifuge and fixation without acetic acid in 80 % methanol is effective for detecting colocalization of entromeric alpha-satellite DNA sequences with the kinetochore CENP-B proteins (Marcais et al. 1999). Such FISH analyses of chromosome 21-specific alphoid DNA and immunostaining of kinetochores on extended interphase chromatin fibers and interphase nuclei indicated that centromeric kinetochore-specific proteins bind to restricted areas of centromeric DNA arrays. In general application, this approach allows prevention of protein and DNA loss during processing cell suspensions for cytogenetic and immunochemical evaluation. Immuno-FISH is found to be applicable in cancer research/diagnosis (immunophenotyping during single-cell genetic analysis), studies of chromosome structure and organization, transplantation research, and identification of supramolecular disease mechanisms (Meaburn et al. 2009; Strickfaden et al. 2010). Figure 11.6 shows immuno-FISH on interphase neuronal cells of the adult human brain (more details in Iourov et al. 2009b, c).

Fast-FISH with microwave activation for I-FISH: Usually FISH using chromosome-specific or site-specific DNA probes is performed during 1 or 2 days. Several fast-FISH protocols were developed using microwave activation for rapid hybridization and detection. Microwave activation for FISH has been proposed by Dr. Ilia Soloviev in 1994. In contrast to standard FISH protocols, this method offers an opportunity to detect hybridization signals within a few minutes, granting 10- to 15-fold detection time reduction. No signal amplification is used to minimize the overlapping and nonspecific background of hybridization signals during chromosomal analysis of interphase nuclei. Microwave activation makes FISH applicable to cells containing cytoplasm (Fig. 11.7). This technique was highly reproducible and applicable for

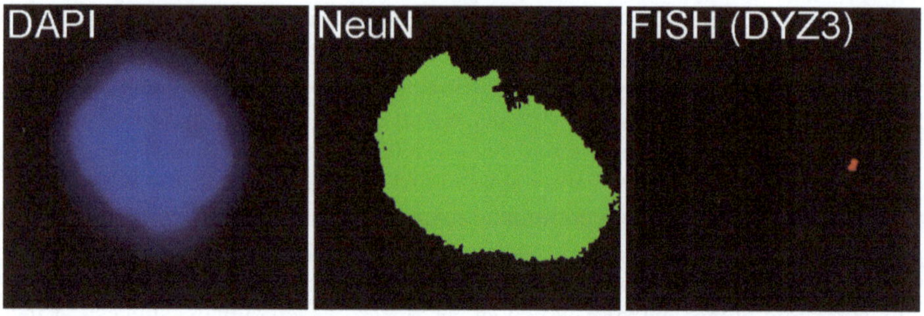

Fig. 11.6 Immuno-FISH (I-FISH) using centromeric probe for chromosome Y (DYZ3) with immunostaining by NeuN (neuron-specific antibody) performed for the analysis of cells isolated from the human brain. (Copyright © Vorsanova et al. 2010a; licensee BioMed Central Ltd. This is an Open Access article distributed under the terms of the Creative Commons Attribution License (http://creativecommons.org/licenses/by/2.0)

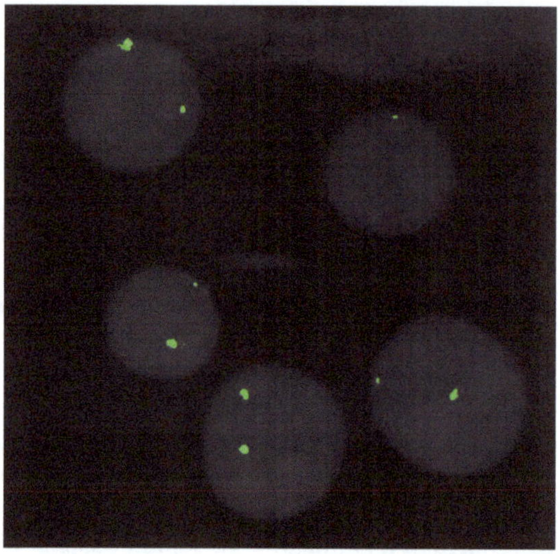

Fig. 11.7 I-FISH performed with microwave activation. FISH signals show centromeric DNAs of chromosomes X and Y (karyotype: 46, XY)

many different chromosome-specific DNA probes. For instance, we have tested alpha-satellite DNA probes specific for chromosomes 1, 3–19, 21, 22, X, and Y. The procedure has allowed rapid chromosome detection in interphase nuclei and metaphase plates of peripheral blood and amniotic fluid (cell suspensions were older than 2 years). Chromosome-specific repetitive DNA probes with FISH microwave activation are to be used for rapid diagnosis of common chromosomal syndromes including chromosome aneuploidies, fast sex determination in prenatal screening, and routine chromosome identification (Soloviev et al. 1994). For more specific purposes, it seems that laboratory microwave ovens are required. However, a common commercially available microwave oven is a handy alternative to a thermal cycler for fast-FISH. Comparable results have been obtained for chromosome

1-/X-specific satellite DNA probes. In addition, the complete fast-FISH procedure was accelerated. An optimized condition for a commercially available X-specific alpha-satellite probe by fast-FISH technique has been also developed for quantitative microscopy (Durm et al. 1997). For highly repetitive DNA probes, the hybridization (renaturation) time and the number of subsequent washing steps can be reduced considerably by omitting denaturing chemical agents (formamide). The appropriate hybridization temperature and time allow a clear discrimination between major and minor binding sites by quantitative fluorescence microscopy. The well-defined physical conditions for hybridization permit automatization of the procedure (Iourov et al. 2008c). Highly fluorescent major binding sites are obtained when denaturation is performed at 74 °C and hybridization is performed during 60 min. These conditions have shown the best microwave activation for denaturation and hybridization to accelerate the procedure. It is to be noted that slides with the target material and the hybridization buffer are placed in a standard microwave oven. After denaturation for 20 s at 900 W, hybridization is performed for 4 min at 90 W. The suitability of a microwave oven for fast-FISH was confirmed with a chromosome 1-specific alpha-satellite probe. In this series of tests, denaturation was performed at 630 W for 60 s and hybridization at 90 W for 5 min. The results were analyzed quantitatively and compared to the results obtained by fast-FISH. The major binding sites were clearly discriminated by their brightness (Durm et al. 1997).

Another method for FISH signals enhancing by microwave pulses during DNA–DNA hybridization using a single- or low-copy probe has shown application of microwaves to be effective in diagnostic or research practice because of the enhancement of weak signals. Microwave FISH has been compared systematically with simple FISH protocols, and it was possible to demonstrate that microwave irradiation leads to better FISH results, especially during the first 100 min of hybridization (Weise et al. 2005).

General advantages and limitations: All FISH-based methods require (1) obtaining cells suspensions or performing other biopsy preparations for the analysis, (2) denaturation and hybridization, and (3) microscopic visual/digital analysis of hybridization results (Iourov et al. 2006b; Iourov 2009). The first stage does not cause any complication, when I-FISH is used, because any cell type of a human organism can be processed for such analyses (Iourov et al. 2006b; Vorsanova et al. 2010a). This is the essential advantage of interphase molecular cytogenetics in contrast to classical cytogenetics (metaphase analysis), that is, the ability to analyze chromosomes in all cell types. Regardless of I-FISH limitations (Liehr and Claussen 2002), some modifications such as ICS-MCB allow a view of interphase chromosomes in their integrity. As mentioned earlier and in Chap. 9, ICS-MCB still has some limitations, being, however, the unique way to visualize the whole banded chromosome in a nucleus (Iourov et al. 2007). I-FISH denaturation and hybridization are performed identically to classical FISH-based approaches (Liehr 2009), and no additional drawbacks can be attributed to these procedures. I-FISH microscopic or digital analysis is not associated with any special problem (Iourov et al. 2008c, 2009c). There are also possibilities to apply digital analysis for studying interphase chromosomes: QFISH analysis of signal colocalization (gene fusions: chromosomal translocations in interphase nuclei), ICS-MCB (visualization of chromosomal structures), increasing "signal visibility," and automatic signal detection. Digital analysis is required for multicolor FISH-based assays (SKY, MFISH, multiprobe interphase FISH, or mFISH), which are usually applied to increase the potential of FISH-based assays through simultaneous analysis of multiple targets (Iourov et al. 2009a, b; Lu et al. 2009). A combination of FISH techniques (i.e., mFISH with 2–5 probes, QFISH, and ICS-MCB) has become the basis for an integrated approach toward molecular diagnosis and genome (chromosome) research at supramolecular level in interphase. Usually, the way that FISH results are evaluated (i.e., visual or digital) is determined by features of DNA probes (amount of probes per reaction and DNA sequence affinity) and detection. Consequently, it is better to subdivide I-FISH techniques this way (Table 11.2).

Several general problems of I-FISH application do exist. Differences of hybridization efficiency complicate simultaneous applications of different probe sets (Iourov et al. 2006a). Site-specific probes signals can be overlooked when wcp or centromeric probes are used (because of intensity differences). Probably the simplest solution is the ICS-MCB. However, some interphase FISH protocols with established probe combinations are proven to be effective for diagnostic purposes (Gersen and Keagle 2005; Liehr 2009). DNA replication during the S-phase of the cell cycle is another major source of unusual I-FISH signal appearance. There are recommendations concerning this type of I-FISH artifacts in the available literature, but the analysis can still be hindered by replicative signals. The latter mainly concerns site-specific probes, being, however, observed during I-FISH with centromeric probes, as well (Fig. 11.8). An additional source of artifacts that can be misinterpreted (i.e., considered as false-positive chromosome abnormalities) is the specificity of nuclear genome organization or interphase chromosome architecture. Here, the problem is related to chromosomal associations (Leitch 2000; Iourov et al. 2005; Krueger and Osborne 2006), significantly affecting I-FISH results and becoming even more important when taking into account that numerous cell types are prone to exhibit chromosomal associations/pairing (Fig. 11.8). Such problems are easily managed by QFISH (Iourov et al. 2005) (Fig. 11.8).

I-FISH for Analysis of DNA Replication

I-FISH allows the visualization of replicating genomic DNA sequences in interphase nuclei. FISH has been shown to help discriminate between nonreplicated and replicated regions of the genome in interphase nuclei, based on the number of specific fluorescent signals (Selig et al. 1992). In normal diploid cells, FISH results on nonreplicated DNA are seen as a single signal whereas replicated loci are characterized by doublets (doubling of a signal). The distribution of these two patterns in unsynchronized cell populations can be used to determine the replication time (S phase) of a DNA sequence. The availability of well-mapped genomic probes and the possibility to compare results from different cell lines make this a convenient approach, by which domains of replication timing control mapped at any

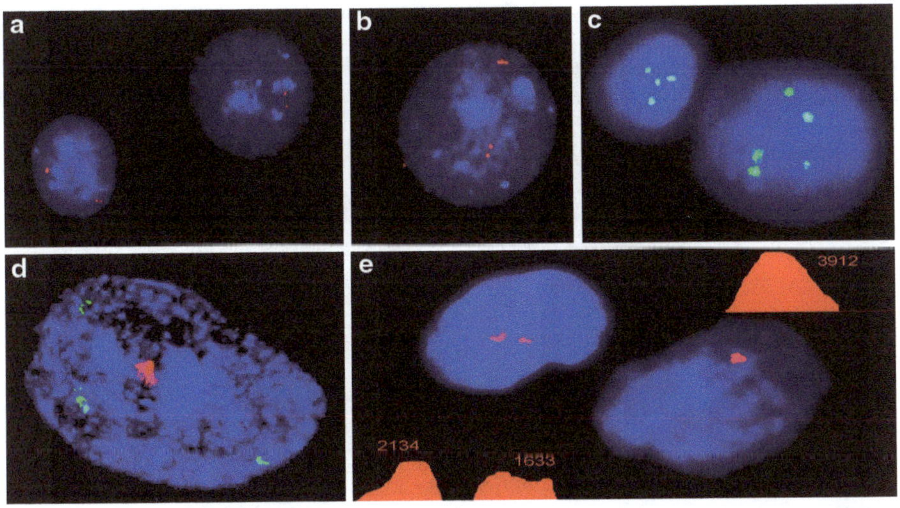

Fig. 11.8 Problems of I-FISH with centromeric/site-specific DNA probes. (**a**, **b**) Replication of specific genomic loci (LSI21 probe): some nuclei exhibit replicated signals, whereas in some nuclei it is not apparent; note the distance between signals can be more than a diameter of a signal. (**c**) Asynchronous replication of a signal (DXZ1) in case of tetrasomy of chromosome X; note difficulty to make a definitive conclusion about number of signals in the right nucleus. (**d**) Two-color FISH with centromeric/site-specific DNA probes for chromosome 1 shows chromosomal associations in a nucleus isolated from the adult human brain; note impossibility to identify number of chromosomes. (**e**) Quantitive FISH (QFISH) demonstrates an association of centromeric regions of homologous chromosomes 9, but not a monosomy or chromosome loss. (Copyright © Vorsanova et al. 2010a; licensee BioMed Central Ltd. This is an Open Access article distributed under the terms of the Creative Commons Attribution License (http://creativecommons.org/licenses/by/2.0)

chromosomal position can be addressed and the relationship to various gene expression patterns can be deduced. Because there appear to be important but poorly understood correlations among replication timing, chromatin structure, and transcriptional competence in mammalian cells, this technique seems to be valuable for understanding related molecular interrelationships.

I-FISH studies have established that monoallelically expressed genes display the unusual property of asynchronous replication, and in those genes that exhibit transcription randomly monoallelic, the asynchronous replication is also random (Ensminger and Chess 2004). By examining the replication timing of genes in a number of human trisomies, authors consistently find one allele replicating early and the other two alleles replicating late, similar to previous observations in X-chromosome trisomies.

I-FISH with chromosome 21-specific cosmid probes was also previously used to identify trisomy 21 in cultured and uncultured amniotic cells. Proper identification of chromosome 21 numbers was made in 65–75 % of trisomic cells and in 70–75 % of normal disomic cells by using all the tested probes. The efficiency of FISH analysis for the total population of interphase cells and cells in the postreplication periods (late S, G_2) of the cell cycle was assessed (Fig. 11.9). Selective scoring of cells in

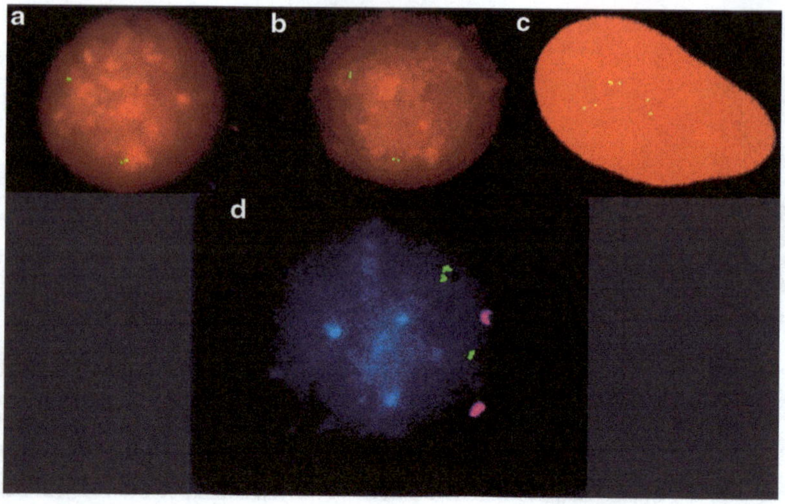

Fig. 11.9 FISH hybridization with the cosmid probe (21q22.2) on cultured amniotic fluid trisomic 21 cells. (**a**) Cell with asynchronously replicating loci (unreplicated locus with singlet signal and replicated locus with doublet signal). (**b**) Replicated cell shows two closely paired hybridization signals corresponding to each chromatid of chromosome 21. (**c**) Cell with trisomy 21 in the postreplicative stage of the cell cycle. (From Soloviev et al. 1995. Reproduced with permission of John Wiley & Sons, Ltd., in the format reuse in a book/textbook vis Copyright Clearance Center). (**d**) Examples of FISH on interphase nuclei with chromosome X-specific centromeric and region-specific probes (locus Xq28) show different types of signals (SD and SD) in a girl with Rett syndrome (RTT). Cy3-labeled centromeric alphoid DNA probe was used. *Two single red signals* indicate simultaneously replicating centromeric DNA from both X chromosomes. PAC clone 671D9 (MeCP2 gene) was labeled by biotin and detected with FITC-avidin. Two asynchronously replicating loci could be seen: *one single green signal* represents late-replicating X chromosome and *one double green signal* represents early-replicating X chromosome. Interphase nuclei were counterstained with fluorescent dye Hoechst 33258 (*blue color*). (From Vorsanova et al. 2001a Brain & development by Nihon Shoni Shinkeigaku. Reproduced with permission of Elsevier BV in the format reuse in a book/textbook via Copyright Clearance Center)

the postreplicative period (a pair of FISH signals on replicated interphase chromosomes) increased the amount of informative nuclei by as much as 95–97 %. The approach was found to determine overlapping chromosomes, artificial doubling of FISH signals on each chromatid of interphase chromosomes, background, and polyploidy. Cosmid probes and integral analysis of hybridization-positive nuclei in pre- and postreplication periods may, therefore, be applicable for improving prenatal diagnosis of trisomy 21. Interestingly, that I-FISH showed that additional chromosomes 21 can induce changing in the replication pattern of an allelic pair: from a synchronous pattern mimicking concomitantly expressed alleles to unsynchronized ones appearing as signals displaying an allele-specific mode of expression (Amiel et al. 1998). A similar phenomenon of asynchronous replication of alleles in genomes carrying an extra chromosome was found in autosomal aneuploidy (trisomy of chromosomes 18 and 13) and sex chromosome aneuploidy (47,XXX and 47,XXY) (Amiel et al. 1999). These data suggest that gross phenotypic abnormalities associated with chromosomal aneuploidy result not only from overexpression of extra gene copies (increased gene dosage) but also from altered expression of genes located on the remaining two homologous chromosomes.

A study of replication timing by I-FISH using chromosome X-specific DNA probes was used to determine the loci with altered replication and transcription in Rett syndrome (RTT), a epigenetic disease caused by mutations in *MECP2*. It was detected that a feature of RTT patients is the *MECP2* locus escaping inactivation in late-replicating chromosome X (Fig. 11.9). Therefore, region Xq28 could contain genes, including *MECP2*, escaping X-inactivation and featured by biallelic expression from the active as from inactive chromosomes X (Vorsanova et al. 2001a). These results support the hypothesis proposing the disturbances in dosage compensation effect caused by aberrant activation of the inactive X-chromosome genes in RTT (biallelic expression in contrast to monoallelic) (Vorsanova et al. 2001a, b) and indicate that normal *M CP2* allele can escape X-inactivation and, in contrast, reduce the pathogenic effect of a mutated allele in RTT.

In the light of the tight relationship between replication timing and expression of a given DNA sequence, the replication timing of *FMR1* alleles on active and inactive X chromosomes was analyzed by I-FISH (Yeshaya et al. 1999). The authors concluded that the *FMR1* locus is subjected to X-inactivation and the delaying effect of the trinucleotide expansion (causing fragile X syndrome) is superimposed on the delay in replication associated with X-inactivation. Thus, a significant epigenetic marker of the interphase chromosome replicative activity is asynchronous replication of monoallelically expressed genes and the synchronous replication of biallelically expressed genes.

Testing a similar hypothesis in microdeletion syndromes (i.e., a microdeletion can affect epigenetic profiling of genes located outside the missing segment), Yeshaya et al. (2009) analyzed the replication patterns of two genes: *SNRPN*, a normally monoallelically expressed gene (assigned to 15q11.13) and *RB1*, a biallelically expressed gene (assigned to 13.q14) in the genomes of patients carrying the 22q11.2 deletion (DiGeorge/velocardiofacial syndrome) and those carrying the 7q11.23 deletion (Williams syndrome). In each affected individual, an aberrant and reversed pattern of replication was shown. In other words, a monoallelic gene replicated more synchronously than a biallelic gene. This inverted pattern, which appears to be nonspecific for those deletions, clearly distinguishes cells of deletion carriers from unaffected individuals. As a result, a potential epigenetic marker for suspecting a hidden microdeletion that is too small to be detected by conventional karyotyping methods was proposed (Fig. 11.10).

Litmanovitch et al. (1998) have used I-FISH for studying replication patterns of alpha-satellite DNA sequences in the light of the human centromere structure and function. They showed an association between replication timing of alpha-satellite DNA sequences and centromere function. Chromosomes having homologous alpha-satellite loci, which replicated synchronously, were revealed to be associated with a lower occurrence of chromosome-specific aneuploidy, whereas chromosomes exhibiting asynchrony with long intervals between early- and late-replicating loci showed the highest occurrence of chromosome-specific aneuploidy. The latter

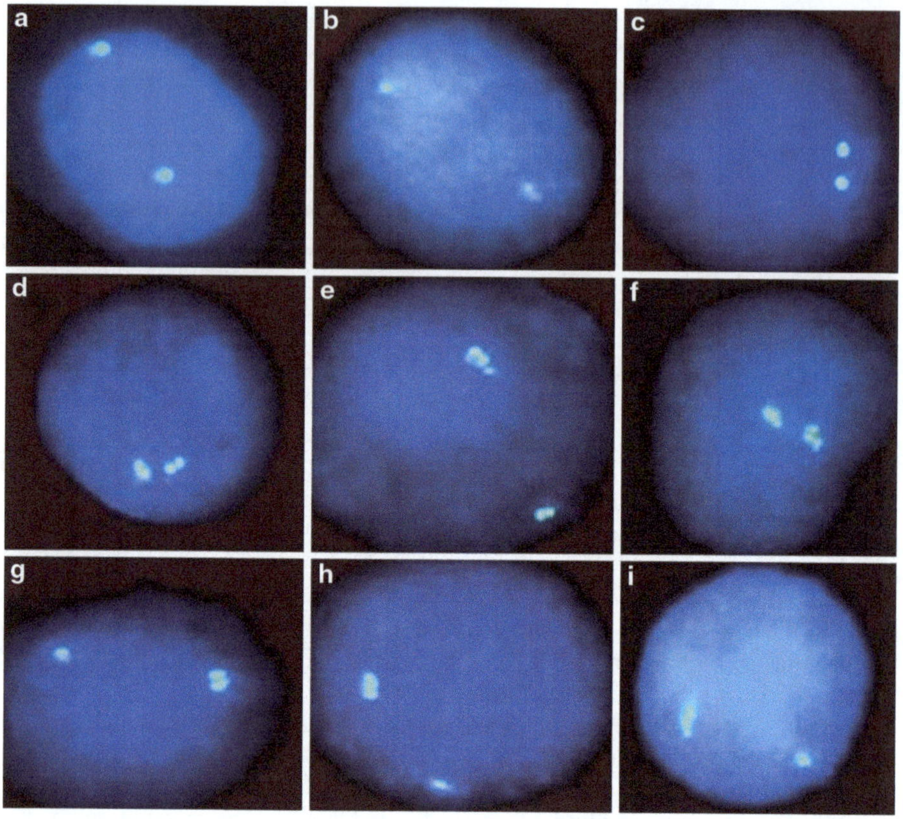

Fig. 11.10 FISH signals in PHA-stimulated lymphocytes at interphase, following FISH with *RB1*. Cells with two singlets (SS cells) in which neither allele has replicated (**a**–**c**); cells with two doublets (DD cells) in which both alleles have replicated (**d**–**f**); and cells with one singlet and one doublet (SD cells) (**g**–**i**), which are S-phase cells in which one allele has replicated while its partner has not. (Copyright © Yeshaya et al. 2009; licensee BioMed Central Ltd. This is an Open Access article distributed under the terms of the Creative Commons Attribution License (http://creativecommons.org/licenses/by/2.0)

supports the hypothesis suggesting that loss of replication control within loc. composed of human centromeric DNAs affects essential centromere functions, such as ensuring proper sister chromatid separation and proper chromosomal segregation during cell division.

Chromosome Architecture and Behavior in Interphase

Chromosome architecture in interphase is consistently shown to be a driving force for crucial intranuclear processes. Specific arrangement of interphase chromosomes is likely to play a role in the regulation of genome activity and cell division as well as formation of chromosome rearrangements occurring during meiosis and mitosis (Leitch 2000; Iourov 2012). To analyze genome organization in interphase, numerous approaches are to be applied, among which I-FISH appears to play the leading role. Several applications of I-FISH approaches for interphase chromosome analysis can be proposed: (1) identification of chromosome positioning and its relation to other nuclear compartments/structures (I-FISH with wcp, interphase MFISH, or ICS-MCB); (2) analyzing positioning of specific genomic loci in relation to each other (associations of whole chromosomes or chromosomal loci) and their behavior (transcriptional/replicative activity) to get a view of functional nuclear genome/chromosome organization and its driving forces (I-FISH with centromeric, site-specific, and wcp, mFISH/QIFSH, or ICS-MCB); and (3) assessment of chromosome arrangement or behavior and its relationship to genome, epigenome, and proteome profiling for delineation of possible consequences of specific interphase chromosome architecture (somatic chromosomal mutations) (I-FISH with centromeric, site-specific, and wcp, mFISH/QIFSH, ICS-MCB, and immuno-FISH). I-FISH analysis of spatial chromosome organization is also influenced by specificity of methods used for structural preservation of nuclei. There are some reports about dependence of fixation on I-FISH results, whereas other studies have not provided similar data. Suspension FISH (S-FISH) is likely to be an alternative for I-FISH spatial genome analysis using standard fixation protocols and is able to leave aside related problems (Steinhaeuser et al. 2002). This technique is discussed in detail in Chap. 10 of this book. In brief, advantages of this approach are referred to the possibility of studying three-dimensionally (3D) preserved nuclei from any human tissue, whereas other 3D preservation techniques require specific conditions of cell cultivation. The latter makes I-FISH lose its main advantage—the opportunity to analyze nondividing cells.

Molecular Cytogenetic Diagnosis

Molecular cytogenetic identification of chromosomal aberrations by I-FISH has been already mentioned in this chapter as well as in a number of comprehensive reviews (Leitch 2000; Iourov et al. 2006a, 2009c; Yurov et al. 2009; Sen and Hopwood 2010; Vorsanova et al. 2010a, b). However, some additional comments about more specific problems of medical cytogenetics seem to be required. Because studying chromosomes in interphase nuclei has undoubtedly profound effects on molecular cancer and prenatal diagnosis, it is obvious that it is impossible to refer all the I-FISH diagnostic studies. To list some technical solutions in molecular cytogenetic diagnosis by I-FISH, we have preferred to focus on difficulties encountered during the introduction and usage for diagnostic purposes. Newly introduced interphase techniques are primarily used for research purposes and are rarely tested for diagnostic validity. Limiting practical application of such I-FISH protocols requires reevaluating the drawbacks. However, the majority of these can be eliminated by application of additional FISH-based approaches (i.e., QFISH). Another problem comes from the diagnosis of somatic chromosomal mosaicism. Regardless of some attempts to propose guidelines or criteria for mosaicism definition (for details, see Iourov et al. 2009c; Vorsanova et al. 2010b), additional studies of somatic mosaicism seem to be strongly required. For instance, a large-scale study aimed to uncover somatic genomic variations in several unaffected human tissues might lead the way. Finally, it is still poorly understood whether data obtained through interphase analysis can be more valid than those obtained by metaphase analysis. From the "structural point of view" (analyzing structural chromosome imbalances), metaphase chromosomal analysis is likely to be more precise. On the other hand, mosaics require large cell populations to be analyzed, and this problem is even more notable when cases of complex, hidden (cryptic), or dynamic mosaicism are evaluated. Metaphase analysis in these cases can be applied for thorough definition of all cell lines, because simple I-FISH analyses (apart from ICS-MCB) are hardly able to show precisely the structure of rearranged chromosomes in a given cell line. More sophisticated studies can require additional data to obtain. For instance, parental origin of chromosomes or epigenetic features addressed by either QFISH (Iourov et al. 2005) or pod-FISH (Weise et al. 2010) could be useful for more thorough confirmational or exclusive diagnosis.

Molecular cytogenetic diagnosis should be performed using a panel of FISH-based techniques (Liehr 2009; Bridger and Volpi 2010; Vorsanova et al. 2010b). To achieve the highest resolution, one can combine molecular cytogenetic techniques based on different platforms (array CGH with I-FISH; metaphase FISH-based techniques with I-FISH, etc.). Cases of complex mosaics or balanced structural chromosome abnormalities seem to especially require such a complex diagnostic procedure. Consequently, regardless of significant developments in molecular interphase cytogenetics, I-FISH techniques remain an addition to whole-genome screening approaches based on array CGH (array CHG) and/or metaphase cytogenetic analysis used for the diagnosis. Only a few targeted I-FISH assays for identification of known caner-associated translocations in interphase and preimplantation genetic diagnosis seem to be applicable in routine molecular cytogenetic diagnosis. To this end, the diagnostic potential of I-FISH is to be more thoroughly analyzed for becoming a routine testing procedure in molecular diagnosis.

Conclusion

According to the present overview of molecular cytogenetic techniques for visualizing chromosomes in interphase, we conclude that a firm technological basis does exist for high-resolution analyses of chromosomes in almost all human tissues. I-FISH advanced by developments in interphase molecular cytogenetics is almost the unique technological issue for studying functional consequences of spatiotemporal chromosome arrangement (architecture) in the interphase nuclei, elucidating the role of such immense intercellular genomic diversity or somatic genomic variations (somatic mosaicism), and proposing new diagnostic solutions for medical genetics, reproductive medicine, and oncology. I-FISH provides for assessment of genome variations and behavior (including DNA replication) in all the cell types of the human organism (all stages of the cell cycle) at molecular resolutions. The combinations of interphase molecular cytogenetic techniques (i.e., mFISH, QFISH, ICS-MCB, S-FISH, pod-FISH, immuno-FISH, etc.) have already given rise to several biomedical discoveries or even new biomedical directions (i.e., molecular neurocytogenetics; for details, see Chap. 3). Therefore, one can insist that developments in interphase molecular cytogenetics are promising for basic and diagnostic research in genetics, cellular and molecular biology, and molecular (genome) medicine. In summary, describing the technological solutions for studying human interphase chromosomes allows us to conclude that interphase molecular cytogenetics opens new opportunities for genetics and cell biology.

Acknowledgments The authors are supported by DLR/BMBF (RUS 2011–2013) and RFFI grant program 12-04-00215- (Russian Federation, 2012–2014). We gratefully acknowledge the NICHD Brain and Tissue Bank for Developmental Disorders at the University of Maryland, Baltimore, MD, USA for providing the brain tissue samples for I-FISH experiments (partially shown in Figs. 11.1, 11.3, 11.4, 11.5, 11.6, and 11.8). Dr. IY Iourov is supported by the Grant of the President of the Russian Federation MD-4401.2013.7.

References

Amiel A, Avivi L, Gaber E, Fejgin MD (1998) Asynchronous replication of allelic loci in Down syndrome. Eur J Hum Genet 6(4):359–364

Amiel A, Korenstein A, Gaber E, Avivi L (1999) Asynchronous replication of alleles in genomes carrying an extra autosome. Eur J Hum Genet 7(2):223–230

Aubert G, Hills M, Lansdorp PM (2012) Telomere length measurement-caveats and a critical assessment of the available technologies and tools. Mutat Res 730(1–2):59–67

Baumgartner A, Weier JF, Weier H-UG (2006) Chromosome-specific DNA repeat probes. J Histochem Cytochem 54:1363–1370

Bejjani BA, Shaffer LG (2008) Clinical utility of contemporary molecular cytogenetics. Annu Rev Genomics Hum Genet 9:71–86

Bridger JM, Volpi EV (eds) (2010) Fluorescence in situ hybridization (FISH): protocols and applications. Humana Press, New York

Carter NP (2007) Methods and strategies for analyzing copy number variation using DNA microarrays. Nat Genet 39:S16–S21

Cremer T, Landegent J, Brückner A, Scholl HP, Schardin M, Hager HD et al (1986) Detection of chromosome aberrations in the human interphase nucleus by visualization of specific target DNAs with radioactive and non-radioactive in situ hybridization techniques: diagnosis of trisomy 18 with probe L1.84. Hum Genet 74(4):346–352

Dundas SR, Boyle S, Bellamy CO, Hawkins W, Garden OJ, Ross JA, Bickmore W (2001) Dual Y-chromosome painting and immunofluorescence staining of archival human liver transplant biopsies. J Histochem Cytochem 49:1321–1322

Durm M, Haar F-M, Hausmann M, Ludwig H, Cremer C (1997) Optimized Fast-FISH with α-satellite probes: acceleration by microwave activation. Braz J Med Biol Res 30(1):15–22

Ensminger AW, Chess A (2004) Coordinated replication timing of monoallelically expressed genes along human autosomes. Hum Mol Genet 13:651–658

Fung J, Weier H-UG, Pedersen RA (2001) Detection of structural and numerical chromosome abnormalities in interphase cells using spectral imaging. J Histochem Cytochem 49:797–798

Gersen SL, Keagle MB (2005) The principles of clinical cytogenetics, 2nd edn. Humana Press, Totowa, NJ

Goetze S, Mateos-Langerak J, van Driel R (2007) Three-dimensional genome organization in interphase and its relation to genome function. Semin Cell Dev Biol 18(5):707–714

Halder A, Jain M, Kabra M, Gupta N (2008) Mosaic 22q11.2 microdeletion syndrome: diagnosis and clinical manifestations of two cases. Mol Cytogenet 1:18

Heng HH, Squire J, Tsui LC (1992) High-resolution mapping of mammalian genes by in situ hybridization to free chromatin. Proc Natl Acad Sci U S A 89(20):9509–9513

Iourov IY (2009) Microscopy and imaging systems. In: Liehr T (ed) Fluorescence in situ hybridization (FISH)—application guide. Springer Verlag, Berlin, pp 75–84

Iourov IY (2012) To see an interphase chromosome or: how a disease can be associated with specific nuclear genome organization. BioDiscovery 4:5. doi:10.7750/BioDiscovery.2012.4.5

Iourov IY, Soloviev IV, Vorsanova SG, Monakhov VV, Yurov YB (2005) An approach for quantitative assessment of fluorescence in situ hybridization (FISH) signals for applied human molecular cytogenetics. J Histochem Cytochem 53:401–408

Iourov IY, Vorsanova SG, Yurov YB (2006a) Chromosomal variations in mammalian neuronal cells: known facts and attractive hypotheses. Int Rev Cytol 249:143–191

Iourov IY, Vorsanova SG, Pellestor F, Yurov YB (2006b) Brain tissue preparations for chromosomal PRINS labeling. Methods Mol Biol 334:123–132

Iourov IY, Vorsanova SG, Yurov YB (2006c) Intercellular genomic (chromosomal) variations resulting in somatic mosaicism: mechanisms and consequences. Curr Genomics 7:435–446

Iourov IY, Liehr T, Vorsanova SG, Yurov YB (2007) Interphase chromosome-specific multicolor banding (ICS-MCB): a new tool for analysis of interphase chromosomes in their integrity. Biomol Eng 24:415–417

Iourov IY, Vorsanova SG, Yurov YB (2008a) Recent patents on molecular cytogenetics. Recent Pat DNA Gene Seq 2:6–15

Iourov IY, Vorsanova SG, Yurov YB (2008b) Molecular cytogenetics and cytogenomics of brain diseases. Curr Genomics 9:452–465

Iourov IY, Vorsanova SG, Yurov YB (2008c) Fluorescence intensity profiles of in situ hybridization signals depict genome architecture within human interphase nuclei. Tsitol Genet 42(5):3–8

Iourov IY, Vorsanova SG, Liehr T, Yurov YB (2009a) Aneuploidy in the normal, Alzheimer's disease and ataxia-telangiectasia brain: differential expression and pathological meaning. Neurobiol Dis 34:212–220

Iourov IY, Vorsanova SG, Liehr T, Kolotii AD, Yurov YB (2009b) Increased chromosome instability dramatically disrupts neural genome integrity and mediates cerebellar degeneration in the ataxia-telangiectasia brain. Hum Mol Genet 18:2656–2669

Iourov IY, Vorsanova SG, Soloviev IV, Yurov YB (2009c) Interphase FISH: detection of intercellular genomic variations and somatic chromosomal mosaicism. In: Liehr T (ed) Fluorescence in situ hybridization (FISH)—application guide. Springer, Berlin, pp 301–311

Iourov IY, Vorsanova SG, Yurov YB (2010) Somatic genome variations in health and disease. Curr Genomics 11(6):387–396

Iourov IY, Vorsanova SG, Yurov YB (2012) Single cell genomics of the brain: focus on neuronal diversity and neuropsychiatric diseases. Curr Genomics 13(6):477–488

Klewes L, Höbsch C, Katzir N, Rourke D, Garini Y, Mai S (2011) Novel automated three-dimensional genome scanning based on the nuclear architecture of telomeres. Cytometry A 79(2):159–166

Krueger C, Osborne CS (2006) Raising the curtains on interchromosomal interactions. Trends Genet 22:637–639

Lee C, Wevrick R, Fisher RB, Ferguson-Smith MA, Lin CC (1997) Human centromeric DNAs. Hum Genet 100:291–304

Leitch AR (2000) Higher levels of organization in the interphase nucleus of cycling and differentiated cells. Microbiol Mol Biol Rev 64(1):138–152

Levsky JM, Singer RH (2003) Fluorescence in situ hybridization: past, present and future. J Cell Sci 116(pt 14):2833–2838
Liehr T (2009) Fluorescence in situ hybridization (FISH)—application guide. Springer, Berlin
Liehr T, Claussen U (2002) Multicolor-FISH approaches for the characterization of human chromosomes in clinical genetics and tumor cytogenetics. Curr Genomics 3:231–235
Liehr T, Pfeiffer RA, Trautmann U, Gebhart E (1998) Centromeric alphoid DNA heteromorphisms of chromosome 22 as revealed by FISH-technique. Clin Genet 53:231–232
Liehr T, Heller A, Starke H, Rubtsov N, Trifonov V, Mrasek K, Weise A, Kuechler A, Claussen U (2002) Microdissection based high resolution multicolor banding for all 24 human chromosomes. Int J Mol Med 9:335–339
Liehr T, Starke H, Weise A, Lehrer H, Claussen U (2004) Multicolor FISH probe sets and their applications. Histol Histopathol 19:229–237
Litmanovitch T, Altaras MM, Dotan A, Avivi L (1998) Asynchronous replication of homologous alpha-satellite DNA loci in man is associated with nondisjunction. Cytogenet Cell Genet 81(1):26–35
Lu CM, Kwan J, Baumgartner A, Weier JF, Wang M, Escudero T, Munne S, Zitzelsberger HF, Weier H-UG (2009) DNA probe pooling for rapid delineation of chromosomal breakpoints. J Histochem Cytochem 57:587–597
Manvelyan M, Hunstig F, Mrasek K, Bhatt S, Pellestor F, Weise A, Liehr T (2008) Position of chromosomes 18, 19, 21 and 22 in 3D-preserved interphase nuclei of human and gorilla and white hand gibbon. Mol Cytogenet 1:9
Marcais B, Vorsanova SG, Roizes G, Yurov YB (1999) Analysis of alphoid DNA variation and kinetochore size in human chromosome 21: evidence against pathological significance of alphoid satellite DNA diminutions. Tsitol Genet 33(1):25–31
Meaburn KJ, Gudla PR, Khan S, Lockett SJ, Misteli T (2009) Disease-specific gene repositioning in breast cancer. J Cell Biol 187:801–812
Mitelman F, Johanson B, Martens F (2007) The impact of translocations and gene fusions on cancer causation. Nat Rev Cancer 7:233–245
Nicholson JM, Duesberg P (2009) On the karyotypic origin and evolution of cancer cells. Cancer Genet Cytogenet 194:96–110
Nietzel A, Rocchi M, Starke H, Heller A, Fiedler W, Wlodarska I et al (2001) A new multicolor-FISH approach for the characterization of marker chromosomes: centromere-specific multicolor-FISH (cenM-FISH). Hum Genet 108:199–204
Raap A, Florijn RJ, Blonden LA, Wiegant J, Vaandrager J-W, Vrolijk H et al (1996) Fiber FISH as a DNA mapping tool. Methods 9(1):67–73
Rouquette J, Cremer C, Cremer T, Fakan S (2010) Functional nuclear architecture studied by microscopy: present and future. Int Rev Cell Mol Biol 282:1–90
Schrock E, du Manoir S, Veldman T, Schoell B, Weinberg J, Ferguson-Smith MA et al (1996) Multicolor spectral karyotyping of human chromosomes. Science 273:494–497
Selig S, Okumura K, Ward DC, Cedar H (1992) Delineation of DNA replication time zones by fluorescence in situ hybridization. EMBO J 11(3):1217–1225
Sen S, Hopwood V (2010) Molecular cytogenetic evidence for multistep tumorigenesis: implications for risk assessment and early detection. Cancer Biomark 9(1–6):113–132
Soloviev IV, Yuri B, Yurov YB, Vorsanova SG, Malet P (1994) Microwave activation of fluorescence in situ hybridization: a novel method for rapid chromosome detection and analysis. Focus 16(4):115–116
Soloviev IV, Yurov YB, Vorsanova SG, Fayet F, Roizes G, Malet P (1995) Prenatal diagnosis of trisomy 21 using interphase fluorescence in situ hybridization of postreplicated cells with site-specific cosmid and cosmid contig probes. Prenat Diagn 15:237–248
Soloviev IV, Yurov YB, Vorsanova SG, Malet P, Zerova TE, Buzhievskaya TI (1998a) Double color in situ hybridization of alpha-satellite chromosome 13, 21 specific cosmid clones for a rapid screening of their specificity. Tsitol Genet 32:60–64
Soloviev IV, Yurov YB, Vorsanova SG, Marcais B, Rogaev EI, Kapanadze BI et al (1998b) Fluorescent in situ hybridization analysis of α-satellite DNA in cosmid libraries specific for human chromosomes 13, 21 and 22. Russ J Genet 34:1247–1255

Speicher MR, Ballard GS, Ward DC (1996) Karyotyping human chromosomes by combinatorial multi-fluor FISH. Nat Genet 12:368–375

Steinhaeuser U, Starke H, Nietzel A, Lindenau J, Ullmann P, Claussen U et al (2002) Suspension (S)-FISH, a new technique for interphase nuclei. J Histochem Cytochem 50:1697–1698

Strickfaden H, Zunhammer A, van Koningsbruggen S, Köhler D, Cremer T (2010) 4D chromatin dynamics in cycling cells: Theodor Boveri's hypotheses revisited. Nucleus 1(3):284–297

Stumm M, Wegner R-D, Bloechle M, Eckel H (2006) Interphase M-FISH applications using commercial probes in prenatal and PGD diagnostics. Cytogenet Genome Res 114:296–301

Vanneste E, Bittman L, Van der Aa N, Voet T, Vermeesch JR (2012) New array approaches to explore single cells genomes. Front Genet 3:44

Verma RS, Luke S (1992) Variation in alphoid DNA sequences escape detection of aneuploidy in interphase FISH technique. Genomics 14:113–116

Virgili A, Brazma D, Reid AG, Howard-Reeves J, Valgañón M, Chanalaris A et al (2008) FISH mapping of Philadelphia negative BCR/ABL1 positive CML. Mol Cytogenet 1:14

Vorsanova SG, Yurov YB, Alexandrov IA, Demidova IA, Mitkevich SP, Tirskaya AF (1986) 18p-syndrome: an unusual case and diagnosis by in situ hybridization with chromosome 18-specific alphoid DNA sequence. Hum Genet 72:185–187

Vorsanova SG, Yurov YB, Deryagin GV, Soloviev IV, Bytenskaya GA (1991) Diagnosis of aneuploidy by in situ hybridization: analysis of interphase nuclei. Bull Exp Biol Med 112:413–415

Vorsanova SG, Yurov YB, Kolotii AD, Soloviev IV (2001a) FISH analysis of replication and transcription of chromosome X loci: new approach for genetic analysis of Rett syndrome. Brain Dev 23:S191–S195

Vorsanova SG, Yurov YB, Ulas VY, Demidova IA, Kolotii AD, Gorbatchevskaia NL, Beresheva AK, Soloviev IV (2001b) Cytogenetic and molecular-cytogenetic studies of Rett syndrome (RTT): a retrospective analysis of a Russian cohort of RTT patients (the investigation of 57 girls and three boys). Brain Dev 23:S196–S201

Vorsanova SG, Yurov YB, Brusquant D, Carles E, Roizes G (2002) Two new cases of the christ-church (Ch1) chromosome 21: evidence for clinical consequences of de novo deletion 21p-. Tsitol Genet 36(1):46–49

Vorsanova SG, Iourov IY, Beresheva AK, Demidova IA, Monakhov VV, Kravets VS et al (2005a) Non-disjunction of chromosome 21, alphoid DNA variation, and sociogenetic features of Down syndrome. Tsitol Genet 39(6):30–36

Vorsanova SG, Kolotii AD, Iourov IY, Monakhov VV, Kirillova EA, Soloviev IV, Yurov YB (2005b) Evidence for high frequency of chromosomal mosaicism in spontaneous abortions revealed by interphase FISH analysis. J Histochem Cytochem 53:375–380

Vorsanova SG, Iourov IY, Voinova-Ulas VY, Weise A, Monakhov VV, Kolotii AD et al (2008) Partial monosomy 7q34-qter and 21pter-q22.13 due to cryptic unbalanced translocation t(7;21) but not monosomy of the whole chromosome 21: a case report plus review of the literature. Mol Cytogenet 1:13

Vorsanova SG, Yurov YB, Iourov IY (2010a) Human interphase chromosomes: a review of available molecular cytogenetic technologies. Mol Cytogenet 3:1

Vorsanova SG, Yurov YB, Soloviev IV, Iourov IY (2010b) Molecular cytogenetic diagnosis and somatic genome variations. Curr Genomics 11:440–446

Walter J, Joffe B, Bolzer A, Albiez H, Benedetti PA, Muller S et al (2006) Towards many colors in FISH on 3D-preserved interphase nuclei. Cytogenet Genome Res 114:367–378

Weier H-UG (2001) DNA Fiber mapping techniques for the assembly of high-resolution physical maps. J Histochem Cytochem 49(8):939–948

Weise A, Liehr T, Claussen U, Halbhuber K-J (2005) Increased efficiency of fluorescence in situ hybridization (FISH) using the microwave. J Histochem Cytochem 53(10):1301–1303

Weise A, Gross M, Hinreiner S, Witthuhn V, Mkrtchyan H, Liehr T (2010) POD-FISH: a new technique for parental origin determination based on copy number variation polymorphism. Methods Mol Biol 659:291–298

Weise A, Mrasek K, Klein E, Mulatinho M, Llerena JC Jr, Hardekopf D et al (2012) Microdeletion and microduplication syndromes. J Histochem Cytochem 60(5):346–358

Wiegant J, Kalle W, Mullenders L, Brookes S, Hoovers JM, Dauwerse JG et al (1992) High-resolution in situ hybridization using DNA halo preparations. Hum Mol Genet 1(8):587–591

Yang F, Shao C, Vedanarayanan V, Ehrlich M (2004) Cytogenetic and immuno-FISH analysis of the 4q subtelomeric region, which is associated with facioscapulohumeral muscular dystrophy. Chromosoma (Berl) 112:350–359

Yeshaya J, Shalgi R, Shohat M, Avivi L (1999) FISH-detected delay in replication timing of mutated FMR1 alleles on both active and inactive X-chromosomes. Hum Genet 105(1–2):86–97

Yeshaya J, Amir I, Rimon A, Freedman J, Shohat M, Avivi L (2009) Microdeletion syndromes disclose replication timing alterations of genes unrelated to the missing DNA. Mol Cytogenet 2:11

Yurov YB, Soloviev IV, Vorsanova SG, Marcais B, Roizes G, Lewis R (1996) High resolution fluorescence in situ hybridization using cyanine and fluorescein dyes: ultra-rapid chromosome detection by directly fluorescently labeled alphoid DNA probes. Hum Genet 97:390–398

Yurov YB, Vorsanova SG, Iourov IY, Demidova IA, Beresheva AK, Kravetz VS et al (2007) Unexplained autism is frequently associated with low-level mosaic aneuploidy. J Med Genet 44(8):521–525

Yurov YB, Vorsanova SG, Iourov IY (2009) GIN 'n' CIN hypothesis of brain aging: deciphering the role of somatic genetic instabilities and neural aneuploidy during ontogeny. Mol Cytogenet 2:23

Appendix: Internet Links for Interphase Molecular Cytogeneticists

Abstract In the final chapter, "Appendix," we have provided a list of Internet sites that are highly useful for those working in the field of interphase molecular cytogenetics. The collection includes sites dedicated to cytogenetics (molecular cytogenetics), clinical aspects of diseases studied by molecular cytogenetic techniques, retrieving scientific (biomedical) literature, and bioinformatic analysis (including resources allowing gene prioritization and pathway analysis, genome browsers, and analyzing tools). The inclusion was performed according to our own experience and reported relevance to genome and chromosome research based on data acquired during molecular cytogenetic analyses. It is clearly impossible to list all the links that are of importance for a scientist whose activity is related to studying genome (chromosome) structure and behavior in interphase. In this instance, we encourage the readers to follow the links provided by the listed resources.

Introduction

To list Internet sites relevant to the field of interphase molecular cytogenetics, we preferred to subdivide the list according to the main modes of application. We also provide references to papers describing the resource (if applicable) and to journals that are publishing papers dedicated to interphase chromosome analyses. However, one has to keep in mind that the provided sites are not all specifically dedicated to interphase cytogenetics. On the other hand, these are almost indispensable to be addressed when a study of genome (chromosome) structure and behavior in interphase at a high technological level is performed. A brief example of the use of in silico (bioinformatic) methods in a study almost completely dedicated to interphase chromosome behavior and variations and its relevance to more general biological processes can be found in Iourov et al. (2009). Finally, these sites are included to the present chapter (Appendix) according to our own experience and reported relevance to genome and chromosome research based on data acquired during molecular cytogenetic analyses. All the links were tested in November 2012.

Y.B. Yurov et al. (eds.), *Human Interphase Chromosomes: Biomedical Aspects*,
DOI 10.1007/978-1-4614-6558-4,

Cytogenetics: Chromosome Abnormalities (Including Cancer Cytogenetic Databases)

arrayMap (genomic arrays for copy number profiling in human cancer) (Cai et al. 2012): http://www.arraymap.org/cgi-bin/amHome.cgi
Atlas of genetics and cytogenetics in oncology and haematology (Huret et al. 2012): http://atlasgeneticsoncology.org/
Chromosomal Variation in Man or Borgaonkar DS. Chromosomal Variation in Man: A Catalog of Chromosomal Variants and Anomalies: Online NLM Version. Bethesda (MD): National Center for Biotechnology Information (US); 1975: http://www.ncbi.nlm.nih.gov/books/NBK105441/ and http://www.wiley.com/legacy/products/subject/life/borgaonkar/access.html
Chromosome Anomaly Collection (managed by Dr John Barber): http://www.ngrl.org.uk/wessex/collection/
CyDAS (drawing derivative chromosomes online) (Hiller et al. 2005): http://www.cydas.org/OnlineAnalysis/
Cytogenetic Gallery (a scholar website for cytogeneticists): http://www.pathology.washington.edu/galleries/Cytogallery/
DECIPHER—database of unbalanced chromosome aberrations (Firth et al. 2009): http://decipher.sanger.ac.uk/
ECARUCA European Cytogeneticists Association Register of Unbalanced Chromosome Aberrations (Feenstra et al. 2006): http://umcecaruca01.extern.umcn.nl:8080/ecaruca/ecaruca.jsp
Mitelman Database of Chromosome Aberrations and Gene Fusions in Cancer: http://cgap.nci.nih.gov/Chromosomes/Mitelman
NCI and NCBI's SKY/M-FISH and CGH Database (2001): http://www.ncbi.nlm.nih.gov/sky/skyweb.cgi
Progenetix (database of cytogenetic abnormalities in cancer) (Baudis 2007): http://www.progenetix.de/progenetix/index.html
Small supernumerary marker chromosomes (sSMC) homepage (managed by Dr. T. Liehr): http://www.med.uni-jena.de/fish/sSMC/00START.htm

Clinical Databases

Autism Chromosome Rearrangement Database (Marshall et al. 2008): http://projects.tcag.ca/autism/
GeneReviews™ or Pagon RA, Bird TD, Dolan CR, et al., editors. GeneReviews™. Seattle, WA: University of Washington, Seattle; 1993: http://www.ncbi.nlm.nih.gov/books/NBK1116/
Genetics Home Reference (consumer-friendly information about the effects of genetic variations on human health): http://ghr.nlm.nih.gov/
MedGen (organizes information related to human medical genetics, such as attributes of conditions with a genetic contribution): http://www.ncbi.nlm.nih.gov/medgen/
OMIM (online Mendelian inheritance in man): http://www.omim.org/
ORPHANET (the portal for rare diseases and orphan drugs): http://www.orpha.net
The Phenotype-Genotype Integrator (PheGenI): http://www.ncbi.nlm.nih.gov/gap/PheGenI
SFARI Gene/AutDB (a publicly available, curated, web-based, searchable database for autism research) (Basu et al. 2009): http://www.mindspec.org/autdb.html
UNIQUE (rare chromosome disorder support group): http://www.rarechromo.org/html/home.asp

FISH and Array CHG-Based Techniques (Including Pages Containing Information About DNA Probes)

A general compilation of links to molecular cytogenetics resources and beyond (the last chapter of the *Fluorescence In Situ Hybridization (FISH): Application Guide*, Edited by Dr. T. Liehr): http://www.springer.com/life+sciences/cell+biology/book/978-3-540-70580-2

ArrayCyGHt (a web-based application tool for analysis and visualization of array-CGH data) (Kim et al. 2005): http://genomics.catholic.ac.kr/arrayCGH/

CHORI BACPAC resources (Children's Hospital & Research Center Oakland; Dr. P. De Jong): http://bacpac.chori.org/about.htm

CCAP BAC Clones (Cancer Genome Anatomy Project, NCI): http://cgap.nci.nih.gov/Chromosomes/CCAP_BAC_Clones

e-FISH (an in-silico FISH simulation tool): http://projects.tcag.ca/efish/

Genomic Clone Database (The Centre for Applied Genomics and The Hospital for Sick Children, Toronto): http://projects.tcag.ca/gcd

The International Standards for Cytogenomic Arrays (ISCA) Consortium: https://www.iscaconsortium.org/

Multicolor FISH database (managed by Dr. T. Liehr): http://www.fish.uniklinikum-jena.de/mFISH.html

NCBI Probe/Probe Database (a public registry of nucleic acid reagents designed for use in a wide variety of biomedical research applications, with information on reagent distributors, probe effectiveness, and computed sequence similarities): http://www.ncbi.nlm.nih.gov/probe

Resources for molecular cytogenetics: http://www.biologia.uniba.it/rmc/

Scitable by Nature Education—Cytogenetics—FISH: http://www.nature.com/scitable/topic/chromosomes-and-cytogenetics-7 http://www.nature.com/scitable/topicpage/fluorescence-in-situ-hybridization-fish-327

University of Kansas Medical Center compilation of links to cytogenetic (molecular cytogenetic) resources: http://www.kumc.edu/gec/prof/cytogene.html

Genome, Epigenome and Pathway Analyzing Tools (Including Genome Browsers, Bioinformatic Tools for Gene Prioritization and Pathway Analysis)

AmiGO browser (Gene Ontology project browser and search engine): http://amigo.geneontology.org/cgi-bin/amigo/go.cgi

BioGPS (a free extensible and customizable gene annotation portal, a complete resource for learning about gene and protein function; The Scripps Research Institute) (Wu et al. 2009): http://biogps.org

A Catalog of Published Genome-Wide Association Studies (NHGRI) (Hindorff et al. 2009): http://www.genome.gov/gwastudies/

ENDEAVOUR (a gene prioritization tool) (Tranchevent et al. 2008): http://homes.esat.kuleuven.be/~bioiuser/endeavour/tool/endeavourweb.php

Ensembl Genome Browser: http://www.ensembl.org/index.html

Gene Expression Omnibus (GEO): http://www.ncbi.nlm.nih.gov/geo/

The Gene Wiki (Wikipedia-based gene annotation portal) (Huss et al. 2010): http://en.wikipedia.org/wiki/Portal:Gene_Wiki

Gene Wanderer (a gene prioritization tool using interactome) (Köhler et al. 2008): http://compbio.charite.de/genewanderer/GeneWanderer

KEGG or Kyoto Encyclopedia of Genes and Genomes (source for understanding high-level functions and utilities of the biological system): http://www.genome.jp/kegg/
NCBI BioSystems Database (access to biological systems and their component genes, proteins, and small molecules, as well as literature describing those biosystems and other related data): http://www.ncbi.nlm.nih.gov/biosystems
NCBI Build 37.1/NCBI Map Viewer (*Homo sapiens*, Annotation Release 104): http://www.ncbi.nlm.nih.gov/projects/mapview/map_search.cgi?taxid=9606
NCBI Gene (a record may include nomenclature, RefSeqs, maps, pathways, variations, phenotypes, and links to genome-, phenotype-, and locus-specific resources worldwide): http://www.ncbi.nlm.nih.gov/gene/
Pathway Commons (maintained by Memorial Sloan-Kettering Cancer Center and the University of Toronto.): http://www.pathwaycommons.org/pc/
REACTOME (manually curated and peer-reviewed pathway database) (Vastrik et al. 2007): www.reactome.org/
UCSC Genome Browser: http://genome.ucsc.edu/
UniHI (interactome analysis tool) (Chaurasia et al. 2007): http://www.unihi.org/

Genomic Variation Databases

Database of Genomic Variants (hosted by the Centre for Applied Genomics): http://dgvbeta.tcag.ca/dgv/app/home?ref=GRCh37/hg19
Database of genomic structural variation (dbVar) http://www.ncbi.nlm.nih.gov/dbvar/
Human Genome Structural Variation Project (Structural Variation Database) (managed by Eichlerlab) http://humanparalogy.gs.washington.edu/structuralvariation/
Human Genome Variation Society: http://www.hgvs.org/dblist/dblist.html

Biomedical Literature (Including Websites Allowing Acquiring Scientific Information Through the Entire World Wide Web)

PubMed: http://www.ncbi.nlm.nih.gov/pubmed
ScienceDirect—Elsevier: (>11 million full-text journal articles and book chapters): http://www.sciencedirect.com/
SCIRUS (for scientific information only; searches more than 545 million science-specific web pages): http://scirus.com/
Scopus (easy and comprehensive resource to support the research needs in the scientific, technical, medical, and social sciences fields): http://www.scopus.com/home.url
Springer (contains nearly six million resources): http://link.springer.com/
Wiley onli ne library: http://onlinelibrary.wiley.com/
HighWire Stanford University: http://highwire.stanford.edu/
BioMed Central The Open Access Publisher: http://www.biomedcentral.com/
Web of Knowledge (Thompson Reuters/impact factor): http://apps.webofknowledge.com
Google Scholar: http://scholar.google.com/
To this end, we found pertinent to provide a list of journals that publish research in interphase molecular cytogenetics. The inclusion was made according to ratio of numbers of interphase cytogenetics papers to the overall number of papers. Journals are sorted by relevance; those journals that have ceased to publish are excluded.
Molecular Cytogenetics: http://www.molecularcytogenetics.org/

Chromosoma: http://link.springer.com/journal/412
Chromosome Research: http://link.springer.com/journal/10577
Journal of Cell Biology: http://jcb.rupress.org/
Journal of Cell Science: http://jcs.biologists.org/
Genes,ChromosomesandCancer:http://onlinelibrary.wiley.com/journal/10.1002/%28ISSN%291098-2264;jsessionid=14F5A0A05993373860006F087881472C.d01t04
Experimental Cell Research: http://www.journals.elsevier.com/experimental-cell-research/
Molecular Biology of the Cell: http://www.molbiolcell.org/
Molecular and Cell Biology: http://mcb.asm.org/
Cytogenetic and Genome Research: http://content.karger.com/ProdukteDB/produkte.asp?Aktion=JournalHome&ProduktNr=224037
Journal of Histochemistry and Cytochemistry: http://jhc.sagepub.com/
Human Genetics: http://www.springer.com/biomed/human+genetics/journal/439
The American Journal of Human Genetics: http://www.cell.com/AJHG/
Human Molecular Genetics: http://hmg.oxfordjournals.org/
BioDiscovery: http://www.biodiscoveryjournal.co.uk/
Current Genomics: http://www.benthamscience.com/cg/
Methods in Molecular Biology: http://www.springer.com/series/7651
PLoS One: http://www.plosone.org/
Cell: http://www.cell.com/home
Nature Cell Biology: www.nature.com/ncb/
Prenatal Diagnosis: http://onlinelibrary.wiley.com/journal/10.1002/%28ISSN%291097-0223
Mutation Research: http://www.elsevier.com/journals/mutation-research-full-set/FS00-0289
JournalofCellularPhysiology:http://onlinelibrary.wiley.com/journal/10.1002/%28ISSN%291097-4652
Journal of Medical Genetics: http://jmg.bmj.com/
Genes & Development: http://genesdev.cshlp.org/
PNAS USA: http://www.pnas.org/
Cancer Research: http://cancerres.aacrjournals.org/

References

Basu SN, Kollu R, Banerjee-Basu S (2009) AutDB: a gene reference resource for autism research. Nucleic Acids Res 37:D832–D836

Baudis M (2007) Genomic imbalances in 5918 malignant epithelial tumors: an explorative meta-analysis of chromosomal CGH data. BMC Cancer 7:226

Cai H, Kumar N, Baudis M (2012) ArrayMap: a reference resource for genomic copy number imbalances in human malignancies. PLoS One 7(5):e36944

Chaurasia G, Iqbal Y, Hänig C, Herzel H, Wanker EE, Futschik ME (2007) UniHI: an entry gate to the human protein interactome. Nucleic Acids Res 35:D590–D594

Feenstra I, Fang J, Koolen DA, Siezen A, Evans C, Winter RM et al (2006) European Cytogeneticists Association Register of Unbalanced Chromosome Aberrations (ECARUCA): an online database for rare chromosome abnormalities. Eur J Med Genet 49(4):279–291

Firth HV, Richards SM, Bevan AP, Clayton S, Corpas M, Rajan D et al (2009) DECIPHER: database of chromosomal imbalance and phenotype in humans using Ensembl resources. Am J Hum Genet 84(4):524–533

Hiller B, Bradtke J, Balz H, Rieder H (2005) CyDAS: a cytogenetic data analysis system. Bioinformatics 21(7):1282–1283

Hindorff LA, Sethupathy P, Junkins HA, Ramos EM, Mehta JP, Collins FS, Manolio TA (2009) Potential etiologic and functional implications of genome-wide association loci for human diseases and traits. Proc Natl Acad Sci U S A 106(23):9362–9367

Huret JL, Ahmad M, Arsaban M, Bernheim A, Cigna J, Desangles F et al (2012) Atlas of genetics and cytogenetics in oncology and haematology in 2013. Nucleic Acids Res 41(D1):D920–D924. doi:10.1093/nar/gks1082

Huss JW 3rd, Lindenbaum P, Martone M, Roberts D, Pizarro A, Valafar F et al (2010) The gene Wiki: community intelligence applied to human gene annotation. Nucleic Acids Res 38:D633–D639

Iourov IY, Vorsanova SG, Liehr T, Kolotii AD, Yurov YB (2009) Increased chromosome instability dramatically disrupts neural genome integrity and mediates cerebellar degeneration in the ataxia-telangiectasia brain. Hum Mol Genet 18(14):2656–2669

Kim SY, Nam SW, Lee SH, Park WS, Yoo NJ, Lee JY, Chung YJ (2005) ArrayCyGHt: a web application for analysis and visualization of array-CGH data. Bioinformatics 21(10):2554–2555

Köhler S, Bauer S, Horn D, Robinson PN (2008) Walking the interactome for prioritization of candidate disease genes. Am J Hum Genet 82(4):949–958

Marshall CR, Noor A, Vincent JB, Lionel AC, Feuk L, Skaug J et al (2008) Structural variation of chromosomes in autism spectrum disorder. Am J Hum Genet 82(2):477–488

Tranchevent LC, Barriot R, Yu S, Van Vooren S, Van Loo P, Coessens B et al (2008) ENDEAVOUR update: a web resource for gene prioritization in multiple species. Nucleic Acids Res 36:W377–W384

Vastrik I, D'Eustachio P, Schmidt E, Gopinath G, Croft D, de Bono B et al (2007) Reactome: a knowledge base of biologic pathways and processes. Genome Biol 8(3):R39

Wu C, Orozco C, Boyer J, Leglise M, Goodale J, Batalov S et al (2009) BioGPS: an extensible and customizable portal for querying and organizing gene annotation resources. Genome Biol 10(11):R130

Index

Y.B. Yurov et al. (eds.), *Human Interphase Chromosomes: Biomedical Aspects*,
DOI 10.1007/978-1-4614-6558-4, © Springer Science+Business Media, LLC 2013

FSC
www.fsc.org
MIX
Papier | Fördert
gute Waldnutzung
FSC® C083411

Kolibri 360 GmbH
Ferdinand-Jühlke-Straße 7
99095 Erfurt, Deutschland
produktsicherheit@kolibri360.de